Cave and Karst Systems of the World

Series editor

James W. LaMoreaux, Tuscaloosa, AL, USA

For further volumes:
http://www.springer.com/series/11987

Stefan Shanov • Konstantin Kostov

Dynamic Tectonics and Karst

Springer

Stefan Shanov
Konstantin Kostov
Department of Seismotectonics, Geological Institute
Bulgarian Academy of Sciences
Sofia
Bulgaria

ISBN 978-3-662-43991-3 ISBN 978-3-662-43992-0 (eBook)
DOI 10.1007/978-3-662-43992-0
Springer Heidelberg New York Dordrecht London

Library of Congress Control Number: 2014945135

Printed on acid-free paper

Springer is part of Springer Science+Business Media (www.springer.com)

Preface

The book *Dynamic Tectonics and Karst* is designed for scientists, professionals, and students, working in the field of karst science, tectonics, geomorphology, and paleoseismology. It tries to systematize the gained knowledge about the relationship between karst and dynamic tectonics and to present new approaches to the study of this relationship.

The structure of the book is summarized in the following chapters:

- Chapter 1 is an introductory chapter. It outlines the history of the accomplished studies dealing with the relationship "karst and tectonics." This relationship was mentioned even in the works of the first karstologists at the end of the nineteenth century. During the last three decades, the study of the dynamic tectonics and recent geodynamics in karst terrains was a theme of different publications but not systematized in a complete monograph. We hope that significant contributions to this field are referred, and especially the less known publications from East European authors will be of interest for the readers.
- Chapter 2 discusses the fundamental notions related to tectonic stress fields and the applicable methods for field studies of the relationship "tectonic stress → fracturing → karst process." Basic theoretical information is given for the electrical anisotropy of rocks and for earthquake mechanisms in the Earth's crust as phenomena that help a lot the correct organization in time and space of the reconstructed tectonic stress fields. The time sequence of the reconstructed stress fields is very important for the understanding of the evolution of the karst process and the correct assessment of the present-day situation in karst areas. A number of case studies from Bulgaria, Albania, Cuba, and France are presented. They give better knowledge of how stress fields control the drainage route of water and the formation of cavities and cave systems. One of the practical aspects of such studies is normally related to the hydrogeology and problems of water supply and water pollution. However, another practical issue in investigating karst systems (especially the active karst) is that it gives knowledge about the youngest, present-day stress field which could be offered as an intelligent and low-costing first approach for planning and control of artificial fracturing in oil and gas fields.
- Chapter 3 is dedicated to methods of study of remains of seismotectonic events in the caves. Karst systems provide a favorable environment to determine the geometrical and mechanical parameters and to date them. Methods of study of the recent geodynamics in karst terrains and the complex methodology are presented. They include analyses of the spatial orientation of deformed speleothems, instrumental measurements and monitoring, mechanical measurements and modeling, and absolute dating of the deformed speleothems. Case studies from

different karst regions in Bulgaria are presented. The important role of the movement itself of the tectonic block during the process of rupturing and displacement along an active fault is denoted. The practical aspects of paleoseismological investigations in the caves are closely related to the problems of seismic hazard evaluation. The number of records of strong earthquakes in the given area is not sufficient for creation of correct linear regression between the number N of earthquakes and their magnitudes M for the given time interval and area. In most cases, the assuring of regression for the strong events can be done only by using the paleoseismological data.

Many people have helped us in numerous ways during the long years of fieldwork. We are especially grateful to the speleologists of clubs "Akademik" and "Edelweiss" of Sofia, as well as from other Bulgarian caving clubs for their friendly support. Without their help in field studies, such a book could be hardly realized. Bulgarian Federation of Speleology initiated several expeditions in different countries and the participation of scientists was one important factor for the successful results. Some of these results are discussed in this book. The studies in France were supported by the speleological club of Saint Herblain.

We thank our colleagues from the Department of Seismotectonics, Department of Hydrogeology and Department of Geological Hazards of Geological Institute of Bulgarian Academy of Sciences for the perfect cooperation and help with valuable information or materials relating to their respective areas of expertise. The team of the Department of Seismology of Royal Observatory of Belgium has important role for initiating the study of paleo-earthquakes in Bulgaria. Especially we thank Serge Delaby from the University of Mons for the practical advices at the beginning of the paleoseismologic studies in the Bulgarian caves. Special thanks also due to Prof. Gyozo Szeidovitz and Dr. Katalin Gribovszki from the Seismological Observatory of the Geodetic and Geophysical Institute in Budapest for the common work in some of the karst areas in Bulgaria. We highly appreciate the comments of Prof. D.Sc. Christo Dabovski in Chaps. 1 and 2.

We express explicitly our appreciation to our families. Without their understanding during our long absences from home and the support when completing the manuscript, this book would not have been possible.

Within the frames of this monograph it is complicating to examine absolutely all the aspects of diversity in the complex relationship between dynamic tectonics and karst. Working on this book was difficult but always exciting for us. Stimulations for future studies are what we fervently hope our readers will gain from this modest effort.

Sofia, September 2013

Stefan Shanov
Konstantin Kostov

Acknowledgments

We are grateful to the following publishers for kind permission to use their copyright materials as detailed below. Credit to the authors is also given in the figure captions and table headings.

Geologica Balcanica, Geological Institute, BAS, Sofia

Shanov S (1996) Young tectonics and karst formation in the Albanian Alps. Geologica Balcanica, Sofia, 26(3):47–52

Bulletin of Bulgarian Geological Society, Geological Institute, BAS and Bulgarian Geological Society, Sofia

Paskalev M, Benderev A, Shanov S (1992) Tectonic conditions of the region of the Iskrets Karst Springs (West Stara-Planina). Rev Bulg Geol Soc, vol 53(2):69–81

Bulgarian Geophysical Journal, National Institute of Geophysics, Geodesy and Geography, BAS, Sofia

Georgiev Tz, Shanov S (1991) Contemporary geodynamics of the western part of the Moesian Platform (Lom Depression). Bulg Geoph J LXVII, No 3:3–9

Contents

Symbols

C	Parallel electrical conductivity
f_n	Normal strain
h	Layer thickness
I	Electrical current
K	Uniaxial tensile strength
P	Tectonic compression
R	Electrical resistance
T	Tectonic tension
TR	Transversal electrical resistance
ΔU	Electrical potential difference
θ	Shear angle
λ	Coefficient of anisotropy
μ	Coefficient of internal friction
μ_e	Frictional coefficient along the walls of the initial joints
$\boldsymbol{\nu}$	Coefficient of Poisson
ρ	Electrical resistivity
ρ_a	Apparent resistivity (measured in situ resistivity)
ρ_l	Average parallel electrical resistivity
ρ_t	Average transversal electrical resistivity
σ_1	Maximal principal tectonic stress
σ_2	Intermediate principal tectonic stress
σ_3	Minimal principal tectonic stress
σ_φ	Normal stress on the critical plane
τ_c	Cohesion shear strength
τ_f	Frictional resistance
τ_φ	Tangential stress acting on the plane oriented at the angle φ towards the maximum stress σ_1
φ	Direction of maximum stress concentration

Tectonic Control on Karst Evolution

1

1.1 State of Knowledge

The relationship between karst and tectonics is briefly mentioned in the books by the pioneers of karst science: E.A. Martel, Jovan Cvijic, and others. The structural control on the formation of karst cavities was described by one of the founders of speleology Martel in his book "Les Abimes." Martel thought that principally all vertical caves are giant marmites, formed by the mechanical and chemical influence of waters, swallowed by faults and fractures (Martel 1894).

The tectonic preconditions of karstification have been also discussed in the 47 fundamental works of Cvijic, dedicated to the karst geomorphology and hydrology: "The uplift of some limestone surface remote the base level presented by impermeable strata or zone fill of water resulting in very intensive development of the karst erosion. The lowering of the limestone surface causes retroactivity: the zone filled with water is closer to the surface and the karst erosion is weakened." (Cvijic 1989, p. 181).

The connection between vertical tectonic movements and the development of karst is also described in the monograph of the Bulgarian geomorphologist Zheko Radev: "…Together with the gradual uplifting of the mountain were changed the dispositions of the impermeable strata which formed the base of the karst water. After time this circumstance caused the mechanical erosion transfer and creating of new evolutionary levels with remodelling of the karst terrain…": (Radev 1915, p. 130).

The tectonic control on the formation of surface and underground karst landforms is described in the literature mostly as an influence of the tectonic structures (anticlines, synclines, faults, fractures) and (or) vertical crustal movements as a precondition to development of leveled karst (Deike1969; Glazek 1989; Choppy 1997; Ekmekci 2003; Yan et al. 2008). The tectonic control on the formation of karstic landforms as karren, dolines, sinkholes, poljes, etc., has been also discussed for different geographic territories (ex. Renault 1968; Habic 1982; Čar 1982, 1985; Bitencourt Rodet 2001; Faivre and Reiffsteck 2002; Popit 2004, and others).

During the last three decades, the study of the dynamic tectonics and recent geodynamics in karst terrains was a theme of different publications but not systematized in a complete monograph. The Symposium "Karst and Tectonics," held at the cave site of Han-sur-Lesse, Belgium in March 1998, brought together researchers from Belgium, Canada, Czech Republic, France, Great Britain, Israel, Italy, Luxembourg, Portugal, Romania, Spain, and Switzerland to report their studies on the interaction between karst and tectonics. The most important results of this meeting were published in *Geodynamica Acta 12 n° 3–4* (13 papers—Benkovics et al. 1999; Boinet 1999; Bonnet and Colbeaux 1999; Crispim 1999; Devos et al. 1999; Fenard et al. 1999; Gilli 1999; Gilli and Delange 1999; Gilli et al. 1999; Lemeille et al. 1999; Quinif 1999; Rodet 1999; Thery et al. 1999), *Geodynamica Acta 12 n° 12* (three papers—Bracq and Brunin 1999; Hauselmann et al. 1999; Pereira et al. 1999) in 1999, *Geodinamica Acta 13 n^o* 5 in 2000 (two papers—Astruc et al. 2000; Tognini 2000), and one paper in *Bull. Centre Rech. Elf Explo. Prod.* in 2000 (Montenat et al. 2000).

The "red line" in these papers is that karstification is a result of tectonics. It was demonstrated in examples from different areas that karst terrains are

S. Shanov and K. Kostov, *Dynamic Tectonics and Karst*, Cave and Karst Systems of the World,
DOI: 10.1007/978-3-662-43992-0_1,

good recorders of continental deformation in terms of brittle structures and seismic features. Tectonics controls the main directions of karst systems and determines the mechanical limits to the karst expansion. The exact and brief conclusion resulting from these studies is: "No dynamic constraints, no karst !" (Quinif and Vandycke 2001).

All these contributions emphasize the new qualitative view on the relationships between the geodynamical processes and karst evolution, as well as the importance of modern quantitative approaches for deeper understanding of these relations.

The role of tectonic stress fields as a controlling factor of karst and caves development is relatively rarely mentioned in the geological and geomorphological literature. The importance of the paleostress fields as a speleogenetic precondition is described in the works of Jeannin (1990), Choppy (1997), Hauselmann et al. (1999) for selected karst areas in Switzerland and France. The papers published in this period by S. Shanov (Chanov 1988a, b; Shanov and Cousset 1993; Chanov and Benderev 1993; Shanov 1996; Benderev and Shanov 1997; Shanov et al. 2001; Angelova et al. 1999; Shanov and Georgiev 2001) reveal the important role of tectonic stresses as the predestination condition for structuring of karst networks in different terrains in France, Cuba, Albania, and Bulgaria.

Of course, fractures and faults are present as undisputable factors in all works on karst genesis and evolution. Many natural causes exist for the formation of fractures and faults. Factors controlling the occurrence of natural, open, permeable fractures, are the nature and intensity of folding, faulting and in situ stresses acting on rocks with varying in space porosity, bedding, and lithology. The rock properties relating to brittleness are most important for variations in natural fracture density. These properties may vary from regional to local scale; they may vary significantly within the rock formations and between formations. We focus on in situ studies of fracturing and its relationship to stress, especially in rocks potentially capable to be karstified (limestones, dolomites, marbles, etc.). Unfortunately, in publications all over the world focused on problems of karst genesis and development, this very problem is not sufficiently discussed, or the fracture systems are not analyzed from the viewpoint of tectonic stress field.

The neotectonic fracturing in karst regions, especially in regions of recently active karst, is one of the important parameters that control the movement of underground water and the development of karst systems of underground cave passages and potholes. In principle, the most convenient paths for the underground water flow in karstifying massifs are fractures that are oriented perpendicularly to the minimum principal stress. These could be tensile fractures in rock volumes close to the surface, as well as shear joints, formed under conditions of an older stress field, but found to be perpendicular to the minimum compression of the younger tectonic stress field. In such a case, the underground cave passages with active water flow (draining channels) should be oriented perpendicularly to the axis σ_3 of the youngest stress field, or under conditions close to the surface—along the direction of stress σ_1.

The collector underground channel is the main element of the cave system with its own specific individuality. It is connected causally with the lithology, the superimposed tectonic structure, and hydrogeological conditions. Finally, "…it is formed as a result of channel system reorganization, and is a consequence of the presence of a less permeable and less stable fault zone perpendicular to the gradient direction…" (Šušteršič 2003). In brief, since the route of water-penetrating limestone massifs and causing karst phenomena is controlled by tectonics, it is possible to use the inverse route: deduce the style of tectonic movements from the characteristics of karst phenomena.

1.2 Practical Applications

Our long experience in the karst regions in different countries has shown that the karst systems might be a good marker for arranging in the time the reconstructed tectonic stress fields, in particular for recognition of the Neotectonic and Quaternary stress fields, and *vice versa,* the reconstructed and recognized as young stress fields can help the forecasting of the preferential directions of the karst galleries formation and the related dominant flow of the karst waters. According to Vandycke and Quinif (2001) "…karstic caves are favorable sites for the observation and the quantification of recent tectonic activity because they constitute a 3-D framework, protected from erosion…" These authors have shown that a quantitative analysis in term of stress inversion, with the help of striated faults, permitted to reconstruct the stress tensor responsible for the brittle deformation in Rochefort Cave (Namur Province, Belgium). The study of the longest cave in Britain (Farrant and Simms 2011)—Ogof Draenen (70 km), suggests that "…presence of laterally extensive open joints, orientated perpendicular to the regional neo-tectonic principal stress field, determines the depth of flow in the aquifer, rather than fissure frequency."

The linear segments of surface drainages in the present Buffalo River landscape (USA, Northern Arkansas) are parallel to the major joint orientations, and this correlation indicates that erosion and karst dissolution are enhanced by regional jointing (Hudson et al. 2009). According to these authors "…Joints record past stress events, impart fracture permeability, and enhance physical and chemical erosion that contributes to landscape evolution, particularly in karst regions."

Apart from the interest for understanding the tectonic stress field as an important factor for the structured and not chaotic evolution of the karst systems, some purely practical issues have raised when studying the relationship between karst formation and tectonic stress fields. For example, the paper by Karfakis and Loupasakis (2006) presents data about the stress field that caused the main neotectonic structures in the limestone of the Formation Tourkovounia (Greece), the distribution and the geotechnical characteristics of the karstic fissures, the orientation and the geometry of the joint sets, and the geotechnical parameters of the limestones. The paper proves that the influence of karst structures on the geotechnical behavior of the limestones is of significant importance for construction works.

The most unexpected application of the analyses on the fracture systems as a product of the tectonic stress in karst areas is the possibility to determine the preferential directions of increased permeability in oil and gas fields. Especially, this approach is useful when the youngest tectonic stress field is reflected in the features of the contemporary active karst. Of course, the results of the studies on outcrops of karstified carbonate rocks or in the accessible caves near the Earth's surface could not be directly transferred to the depths of the productive oil and gas reservoirs. Using outcrop data to characterize the fracture pattern of a reservoir in depth is not acceptable (Akbar et al. 1993), because the fracture density in reservoirs is commonly lower than values recorded where the same formation outcrops. But, the natural fractures assumed as a product of the regional tectonic stresses and their control on the karst networks, can suggest a preliminary approach when forecasting the preferential direction for inducing fracturing in the reservoirs. Induced fractures are expected to have a preferential orientation subparallel to the natural fractures, as a general tendency, but not obligatory. In such a case, the knowledge of the recent stress field, deduced from the local karst peculiarities, can present valuable practical information.

For example, the Institute of Karst Geology of China (Guilin City, Guangxi Province) promoted in 2011 a Project "Research on the Formation and the Pattern of Carbonate Fracture Cave System." The aim of this Project has been through detailed description and comparative study on the typical karst outcrop area to identify the carbonate fracture cave system in the area of Tahe Oilfield, to characterize Ordovician karst landform units of Tarim oil and gas reservoirs, and to "…launch construction stress field trial in the test area of Tahe Oilfield, starting from the dynamic mechanism of structural cracks, adopting brittle rock fracture criteria, to predict the development condition of tectonic fractures in the test area, to provide theoretical support for the establishment of fracture prediction model in the test area of Tahe Oilfield." (IKG 2011).

Thus, the significance of studies on the relationship between tectonic stress fields, fracturing, and karst formation has not only scientific importance, but is also of promising practical interest.

References

Akbar M, Nurmi R, Standen E, Sharma S, Panwar P, Chaturvedi J, Dennis B (1993) Fractures in the basement: Schlumberger Middle East Well Evaluation Review 14:25–43

Angelova D, Benderev A, Shanov S (1999) Tectonic predestination of the main stages of karst evolution of the Bosnek Karst Region. In: Proceedings of the Jubilee scientific session "half century of systematic conditional geological mapping in Bulgaria", 25–26 Feb 1999, Sofia, pp 3–5 (in Bulgarian)

Astruc JG, Escarguel G, Marandat B, Simon-Coincon R (2000) Floor-age constraining of a tectonic parox-ysm of the Pyrenean orogen. Late middle eocene mammal age of a faulted karstic filling of the Quercy phosphorites, south-western France. Geodin Acta, vol 13, numéro 5:271–280

Benderev A, Shanov S (1997) Karst waters from the region of Bosnek (West Bulgaria): Characteristics and conditions of formation. In: 12th international congress of speleology, La Chaux-de-Fonds, Proceedings, v.2, 6 conference of limestone hydrology and fissured media, pp 255–258

Benkovics L, Obert D, Bergerat F, Mans Y JL, Dubois M (1999) Brittle tectonics and major dextral strike-slip zone in the Budakarst (Budapest, Hungary). Geodin Acta 12 numéro: 201–211

Bitencourt ALV, Rodet J (2001) Premiers éléments d'évolution karstique sous contrôle tectonique d'un massif calcaire : La Serra do Ramalno (Bahia, Brésil). Geologica Belgica, Karst&Tectonics, 4/3–4:251–261

Boinet N (1999) Exploitation de la fracturation d'un massif par la karstification. Exemple du Causse de l'Hortus (Hérault-France). Geodin Acta 12, numéro: 237–247

Bonnet T, Colbeaux JR (1999) L'analyse morphologique spatialisée: apports d'une méthode à la détection des accidents, une nécessité dans l'approche hydrodynamique et karstologique des aquifères fissurés. Exemples dans le Nord de la France crayeux. Geodin Acta 12, numéro 3/4: 223–235

Bracq P, Brunin S (1999) Approche des relations tectonique-karst-hydrodynamics par l'analyse de traçages réalisés dans l'aquifère crayeux du Boulonnais (Escalles, Nord de la France). Geodin Acta 12, numéro 6:359–370

Čar J (1982) Geološka zgradba požiralnega obrobja Planinskega polja (Geological setting of the Planina polje ponor area). Acta Carsologica 10:75–105

Čar J (1985) Geološke osnove oblikovanja kraškega površja. (Geological bases of karst surface formation). Acta Carsologica 14–15:31–38

Chanov S (1988). Conditions géologiques et tectoniques de la formation des cavités karstiques dans la région de "Barkité" et de "Béliar", montagne de Stara Planina près de Vratza / Bulgarie/. Dans "Récit d'une amitié", Saint Herblain, pp 12–14

Chanov S (1988) Contribution à l'étude des champs de tension tectoniques dans le massif karstique d'Arbas /Haute Garonne/.—dans "Récit d'une amitié", Saint Herblain, pp 19–23

Chanov S, Benderev A (1993) Structural, geophysical and hydrogeological exploration on the karst system of Plateau Guaso (Cuba). In: Proceedings of the international symposium on water resources in karst with special emphasis in arid and semi arid zones, 25–26 Oct 1993, Shiraz, Iran, vol 2, pp 595–611

Choppy J (1997) La tectonique et le karst. Proceedings of the 12-th international congress speleology, Switzerland, pp 367–368

Crispim JA (1999) Seismotectonic versus man-induced morphological changes in a cave on the Arrabida chain (Portugal). Geodin Acta 12, numéro 3/4:135–142

Cvijic J (1989) Geography of the limestone terrains. In: Geography of karst (unpublished works of Cvijic J), Collected works, 7, Belgrade. Serbian Academy of Sciences and Arts, pp 125–270 (In Serbian)

Deike HG (1969) Relations of jointing to orientation of solution cavities in limestone of Central Pennsylvania. Am J Sci 267:1230–1248

Devos A, Jaillet S, Gamez P (1999) Structures tectoniques et contraintes de cheminement des eaux dans les aquifères karstiques du Barrais (Lorraine/Champagne—France). Geodin Acta 12, numéro 3/4:249–257

Ekmekci M (2003) Review of Turkish karst with emphasis on tectonic and paleogeographic controls. Acta Carsologica 32/2 17:205–218

Faivre S, Reiffsteck Ph (2002) From doline distribution to tectonic movements: example of the Velebit Mountain range, Croatia. Acta Carsologica 31/3 8:139–154

Farrant AR, Simms MJ (2011) Ogof Draenen: speleogenesis of a hydrological see-saw from the karst of South Wales. Cave Karst Sci 38(1):31–52

Fenard P, Cat NN, Drogue C, Canh DV, Pistre S (1999) Influence of tectonics and neotectonics on the morphogenesis of the peak karst of Halong Bay, Vietnam. Geodin Acta 12, numéro 3/4:193–200

Gilli E (1999) Evidence of palaeoseismicity in a flowstone of the Observatoire cave (Monaco). Geodin Acta 12, numéro 3/4:159–168

Gilli E, Delange P (1999) Recent, slow and aseisic movement of an overthrust observed in the Abel sink hole (Saint-Vallier-de-Thiey, Alpes-Maritimes, France). Geodin Acta 12, numéro 3/4:169–177

Gilli E, Levret A, Sollogoub P, Delange P (1999) Research on the February 18, 1996 earthquake in the caves of Saint-Paul-de-Fenouillet area (earstern Pyrenees, France). Geodin Acta 12, numéro 3/4:143–158

Glazek J (1989) Tectonic conditions for karst origin and preservation. In: Bosak P, Ford D, Glazek J, Horacek I (eds) Paleokarst: a systematic and regional review. Elsevier—Academia, Amserdam—Praha, pp 569–575

Habic P (1982) Karst relief and tectonics (Kraški relief in tektonika, eng. summary). Acta Carsologica 10:26–44

Hauselmann Ph, Jeannin PY, Bitterli T (1999) Relationships between karst and tectonics: case-study of the cave system north of Lake Thun (Bern, Switzerland). Geodin Acta 12, numéro 6:377–387

Hudson MR, Turner KJ, Trexler CC (2009) Regional joint analysis for the Buffalo National River area of Northern Arkansas: implications for tectonic history and landscape influence, 2009 Portland GSA annual meeting (18–21 Oct 2009), Paper No. 136-12

IKG (ed) (2011) Research on the formation and the pattern of carbonate fracture cave system (the first subproject in 973 planning program), Institute of Karst Geology, CAGS, China. http://www.karst.ac.cn/karsten/index.aspx

Jeannin P-Y (1990) Néotectonique dans le karst du nord du Lac de Thoune (Suisse). Karstologia 15:41–54

Karfakis J, Loupasakis C (2006) Geotechnical characteristics of the formation of "Tourkovounia" Limestones and their influence on urban construction—City of Athens, Greece, The Geological Society of London 2006, IAEG 2006, Paper number 794

Lemeille F, Cushing M, Carbon D, Greillet B, Bitterli T, Flehoc C (1999) Co-seismic ruptures and deformations recorded by speleothems in the epicentral zone of the Basel earthquake. Geodin Acta 12, numéro 3/4:179–191

Martel EA (1894) Les Abimes. Delagrave, Paris, p 580

Montenat C, Soudet H, Barrier P, Chereau A (2000) Karstification and tectonic evolution of the Jabal Madar (Adam foothills, Arabian Platform) during the Upper Cretaceous. Bull Centre Rech Elf Explo Prod 22(2):161–190

Pereira E, Rouzaud F, Salottim, Dubois IN, Ferrandini J, Ottavianni-Spella (1999) Castiglione (Oletta, Corsica): relationships between phenomena of calcification and tectonic fossiliferous fracture dating. Geodin Acta 12, numéro 6:371–375

Popit S (2004) Analysis of karst surface landscapes. Int J Fieldwork Stud 2(1). http://www.virtualmontana.org/ejournal/vol2(1)/karst.htm

Quinif Y (1999) Karst et évolution des rivières: le cas de l'Ardenne. Geodin Acta 12, numéro 3/4:267–277

Quinif Y, Vandycke S (2001) "Karst and tectonics"—preface. Geologica Belgica 4(3–4), Karst & Tectonics 171–173. http://popups.ulg.ac.be/Geol/document.php?id=1980

Radev Z (1915) Karst landforms in West Stara Planina. Yearbook of Sofia University, Faculty of history and philology, 10–11. Tzar's Publishing House, p 149 (in Bulgarian)

Renault P (1968) Contribution à l'étude des actions mécaniques et sédimentologiques dans la spéléogénèse. Ann Spéléoi 23(1):259–596

Rodet J (1999) Le réseau de fracturation, facteur initial de la karstification des craies dans les collines du Perche: l'exemple du site de la Mansionnière (Bellou-sur-Huisne, Orne, France). Geodin Acta 12, numéro 3/4:259–265

Shanov S (1996) Young tectonics and karst formation in the Albanian Alps. Geologica Balcanica 26(3):47–52

Shanov S, Benderev A, Ilieva I (2001) Traces of recent active tectonics in the cave of Duhlata (Southwestern Bulgaria). Riviera 2000—Tectonique active et géomorphologie, Villefranche-sur-Mer, Revue d'Analyse Spatiale, No spécial 2001:145–150

Shanov S, Cousset O (1993) Tectonic stress fields and karst processes in the Plateau of Vaucluse (South-Eastern France). In: Proceedings of the international symposium on water resources in karst with special emphasis in arid and semi arid zones, 25–26 Oct 1993, Shiraz, Iran, vol 2, pp 775–783

Shanov S, Georgiev I (2001) Tectonic stresses and their control on the karst formation in the Plateau of Vaucluse (South-East France). Riviera 2000—Tectonique active et géomorphologie, Villefranche-sur-Mer, Revue d'Analyse Spatiale, No spécial 2001, pp 139–144

Šušteršič F (2003) Collapse dolines, deflector faults and collector channels. Speleogenesis Evol Karst Aquifers 1(3):9. Virtual Sci J. www.speleogenesis.info

Thery JM, Pubellier M, Thery B, Butterlin J, Blondeau A (1999) Middle Eocene to Pleistocene example of the Lina Mountains (Irian Jaya, Indonesia). Geodin Acta 12, numéro 3/4:213–221

Tognini P (2000) Role of slumped beds in the genesis and evolution of huge hypogean rooms: an example in the Valle del Nose (Como, Northern Italy). Geodin Acta 13, numéro 5:335–343

Vandycke S, Quinif Y (2001) Recent active faults in Belgian Ardenne revealed in Rochefort Karstic network (Namur Province, Belgium). Neth J Geosci/Geologie en Mijnbouw 80(3–4):297–304

Yan C, Luo G, Wang Y, Chen M, Zhan Q, Wan J, Chen X, Zheng J, Guo J (2008) Compressive structural control on karst development. Sinkholes and the engineering and environmental impacts of karst, pp 54–61. In: 11th multidisciplinary conference on sinkholes and the engineering and environmental impacts of karst, Tallahassee, Florida, United States, 22–26 Sept 2008. doi:10.1061/41003(327)6

2 Tectonic Stress Fields and Karst

2.1 Generalities About the Tectonic Stress Fields

The notion for the *stress* from mechanical point of view was firstly introduced by Augustin Louis Cauchy (1789–1857). He defined the stress as *the average intensity of the strain applied on surface unit.* According to the general sense of the mechanics terminology the stress has to be understood as the integrity of the internal strains that appear when the material is submitted under loads or deformations. Every resulting deformation, no mater of its spatial dimensions, has to be analyzed in the context of the initialing strains. For estimating the state of stress of a given material, the vectors of the impacting stresses on every one of the infinite number of elementary surfaces dividing the material have to be evaluated. Six are the components forming the *stress tensor* inside the elementary volume of given material—three principal and three tangential (Fig. 2.1).

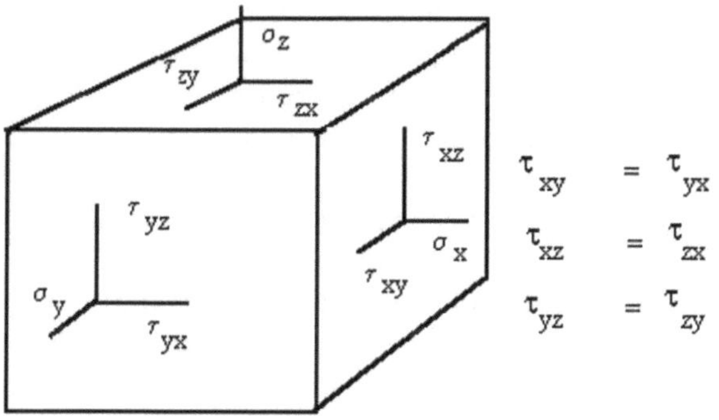

Fig. 2.1 Stress components impacting the elementary volume

No mater how complex could be the applied stresses they can be decomposed at three perpendicularly to each other *normal stresses*. If the values of the applied strains are known, it is possible to evaluate the values and the directions of these three principal stresses. The *stress tensor* defines completely the state of stress. Practically, it is sufficient to know the principal stresses for evaluating the state of stress of the studied material and to determine the values of the normal and tangential stresses on every other plane crossing it. For these purposes, the formulas and the circles of Moore can be used. The method is well described in every book on Engineering Geology and Rock Mechanics (we can suggest the books of Jaeger 1975 or Spenser 1981).

Trollop (1983) introduced the *differential definition of the stress* for expressing the basic conditions of the Theory of Elasticity as continuous mathematical function (Trollop 1983):

$$\sigma = \lim_{\delta_a \to 0} \frac{\delta f_n}{\delta a} = \frac{df}{da} \qquad (2.1.1)$$

where f_n is the normal strain applied on the surface a.

This definition is valuable for ideal continuous medium. But for discrete and granulate materials (that are normally all geological bodies) this differential definition is not acceptable. As it was mentioned above, the total stress state is completely defined by the three principal stresses σ_1, σ_2, and σ_3, where:

$$\sigma_1 > \sigma_2 > \sigma_3. \qquad (2.1.2)$$

The characterization of the state of stress in Tectonophysics is defined at least by three vectors of the principal stresses. If they are equal the state of stress

S. Shanov and K. Kostov, *Dynamic Tectonics and Karst*, Cave and Karst Systems of the World,
DOI: 10.1007/978-3-662-43992-0_2,

is defined as uniform; if not,the state of stress is no uniform. For the uniform state of stress, the values of the stresses are calculated by the values of the applied strains. For the no uniform state of stress the values of the stresses can not be calculate directly, but using the effect of their impact. Exactly, this approach is used for qualitative evaluation of the stress by the different Earth's sciences. The type of the stress field in Tectonophysics is determined by the space orientation of the principal stress axes and their relative rates. Practically, this approach has lead to the definition of the extensional stress (in spite the impossibility to accept this type of mechanical deformation inside the Earth's crust), the compression stress, the strike stress, etc.

Angelier (1994) comments that always when a reconstruction of the stress field is done, one of the reconstructed axes is subvertical. He explains this effect with the gravitational field and the free Earth's surface. The vertical stress is directly dependent from the weight of the rocks and the pore strain. Hence, the tectonic stress is acting, in the most of the cases, on subhorizontal plane. The reconstruction of the paleostresses from the analyses of the fault systems has shown that the deviation of the stress axis determined as a vertical one is surprising small—normally about 10° (Angelier 1994).

The strains creating the deformations in the rocks can be defined in the space by the ellipsoid, the three axes *A*, *B*, and *C* of which determine the maximum, the intermediate, and the minimum strain (Fig. 2.2). Generally, this ellipsoid is randomly space-oriented relatively to the geometric characteristics of the rocks before their deformation. The relationship between the strain ellipsoid and the stress ellipsoid (axes *X*, *Y* and *Z* in Fig. 2.2) is not direct, nor simple. The example in Fig. 2.2 (according to Aubouin et al. 1988) represents the ideal case of isotropic material. Both the two ellipsoids have the same orientation of the axes. The maximum deformation is corresponding to the maximum strain. The general case is the random orientation of the ellipsoids to each other. The field studies on the rock deformations define the orientation of the axes *X*, *Y*, and *Z*. It is not easy, nor simply to determine the principal stresses σ_1, σ_2, and σ_3. But in some conditions regarding the orientation of the principal strains before the deformation, the relationship between the two ellipsoids can be

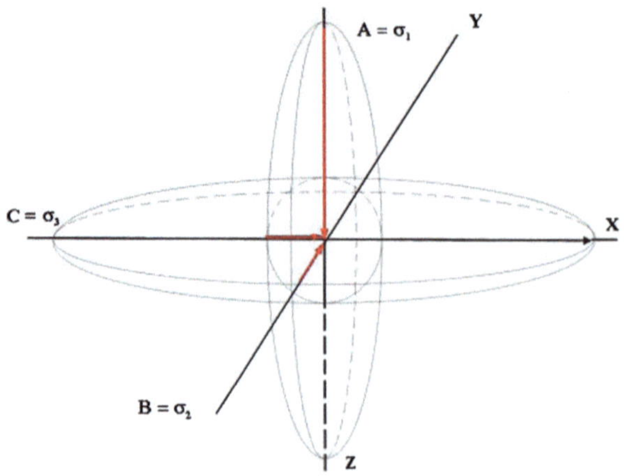

Fig. 2.2 Ellipsoid of the acting strain (*ABC*) and ellipsoid of the deformations (*XYZ*). *A*—maximum compression; *B*—intermediate compression; *C*—minimum compression; *X*—maximum deformation (lengthening); *Y*—intermediate deformation; *Z*—minimum deformation (shortening) (after Aubouin et al. 1988)

simplified. For the structural studies, a simplified acceptance for the relationship *tectonic strain* ⟺ *resulting deformations* in the rocks is applied.

It is well known that the strain and the stress relationship does not directly mean fractures formation, but however strain and stress are causes of fracturing. Hunt et al. (2011) described the principal causes of fracture–density variations (Table 2.1), but concluding immediately that this list of fracture-causing variables is much more extensive, and often unique to a particular fracture system.

As a result of the rock fracturing, different types of single or complex structures appear. The ranging and mutual relationship can be simplified to the scheme presented in Fig. 2.3, following Norsk Geologisk Tidsskrift 69 Supplement (Nystuen 1989).

A fracture is used as a more general term for all kinds of fracturing caused by mechanical stress in the rocks. The term is irrespective of whether dislocation has taken place along the fracture.

A joint is a fracture surface in the rock along which no displacement has occurred. A *joint system* is composed of intersecting joints assumed to have been formed during the same deformational event.

A fault is a fracture separating two bodies of rock, which have moved relatively to one another and it is the fundamental term for a fracture surface in the Earth's crust on which displacement has occurred.

Table 2.1 Causes of fracture-density variation (after Hunt et al. 2011)

Type	Parameter	Direction of correlation
Material property	Rock brittleness	Positive
	Grain size	Negative
	Porosity	Negative
	Bed thickness	Negative
In situ	Depth	Variable effect
	Pore pressure	May hold fractures open
Strain	Structural position	Positive with strain

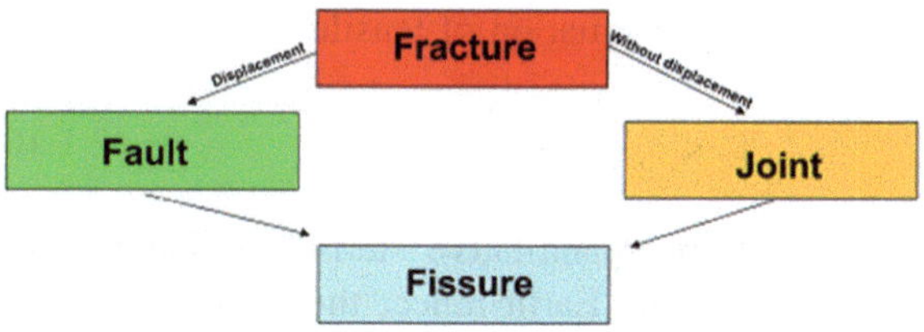

Fig. 2.3 Relationship between the terms fracture, fault, joint, and fissure (according to Norsk Geologisk Tidsskrift 69, Supplement 1989)

A fissure, commonly used term in karst studies, is an open joint or crack. Some fissures appear clearly to be open, but they could be subsequently filled with mineral growth, as well as eroded and enlarged by the water in karstified rocks.

Here a brief overview of the most popular theories of the shear joints formation will be presented, because the shear joint systems are widely used in our studies for reconstruction of the tectonic stress fields' main axes.

One of the most valid theories of failure of rock-like materials is this of Coulomb-Navier, widely accepted in the geotectonic practice. According to this theory, the shear strength consists of two components—cohesion shear strength τ_c and the frictional resistance on the corresponding plane $\tau_f = \mu\sigma_\theta$ (following Stoyanov 1970). The destruction appears when the tangential stress on the plane reaches the value of the strength:

$$\tau_\theta = \tau_c + \mu\sigma_\theta. \qquad (2.1.3)$$

Here θ is the shear angle (the orientation of the shear plane toward the maximum normal stress σ_1) and μ is the coefficient of internal friction.

The orientation of the shear plane is defined by the equation:

$$tg2\theta = \frac{1}{\mu}. \qquad (2.1.4)$$

It follows, that the shear angle θ is depending exclusively from the type of the material. The theory is considered to describe with sufficient accuracy, the destruction of rocks at low and moderate pressures (Brace 1960; Lundborg 1968).

From physical point of view, the Theory of Griffith published in 1921, was accepted as more justified, because it is based on the internal characteristics of the brittle material. This theory accepts that the real material contains a number of randomly oriented micro-joints. Under stress conditions in the material, at the ends of these joints, a stress concentration is present. The maximum stress concentration is along the direction φ. If a critical value of the stress is reached, the micro-joints with orientation φ will grow by secondary micro-fracturing, and finally a macro-fracturing shall be produced. Following this theory McClintock and Walsh (1962) proposed the so-called Modified Griffith's Theory. They have accepted that under a minimum pressure the micro-joints are closed. When differential stresses exist, on the walls of the joints reactive frictional stress appears, being proportional to the compression stress. The equation of destruction is:

$$\tau_\varphi = 2_K + \mu_e\sigma_\varphi. \qquad (2.1.5)$$

Here τ_φ is the tangential stress acting on the plane oriented at the angle φ toward the maximum stress σ_1, K is the uniaxial tensile strength, μ_e is the frictional coefficient along the walls of the initial joints (i.e., it is a coefficient of external friction), and σ_φ is the normal stress on the critical planes (it is positive). Finally, the critical angle of the shear joints is:

$$tg2\varphi = \frac{1}{\mu_e}. \tag{2.1.6}$$

The difference between μ (from Eq. 2.1.4) and μ_e is that the physical meaning of μ is not well defined especially for real rock conditions, while μ_e is a real physical parameter (Stoyanov 1970).

The caves in karst terrains are a result of the water erosion along the most open fractures under given tectonic conditions. In this case, the normal faults and the tensional fractures (fissures) are most favorable for the karst network evolution. The study of the mechanism of formation of fractures in the rocks by Price (1959, 1966) lead to the conclusion, that the vertical tensional and shear fractures and faults are formed as a result of residual elastic stresses during the post-tectonic uplifting of compressed volumes. One very important statement in these works is that these types of structures can be formed only if horizontal tectonic extension exists.

If the Earth's crust is three-dimensionally analyzed, only in the state of the own weight of the rocks, and taking into account that the horizontal enlargement of given volume is not possible because of the environment conditions, σ_x, σ_y and σ_z will be the principal stress axes. It could be written that:

$$\sigma_x = \sigma_y = \frac{\nu}{1-\nu}\sigma_z. \tag{2.1.7}$$

In this case, we accept that the horizontal deformations along the horizontal axes *X* and *Y*, i.e., $\varepsilon_x = \varepsilon_y = 0$, and the axis *Z* is vertical. The coefficient of Poisson ν is less than 0.5 and σ_z will be the *maximum stress = gravitational stress.*

Now it is necessary to add the stress from the tectonic strain. Let us say that the tectonic strain has only horizontal components and the stress effect in every point can be presented by two perpendicular stresses c_x and c_y. The equations of superposition of the horizontal and gravitational stresses will be (after Stoyanov 1970):

$$\begin{aligned} \sigma'_x &= \frac{\nu}{1-\nu}\sigma_z + c_x + \nu c_y \\ \sigma'_y &= \frac{\nu}{1-\nu}\sigma_z + c_y + \nu c_x \\ \sigma'_z &= \sigma_z. \end{aligned} \tag{2.1.8}$$

Our interest in the formation of suitable conditions for the karst evolution is directed to the formation of open fractures. If conditionally, the horizontal stress components $c_y > c_x$, than $\sigma_y > \sigma_x$. The formation of tensile fractures, as well as normal faults in larger scales, can be possible under the condition:

$$\sigma'_x < \sigma'_y < \sigma'_z. \tag{2.1.9}$$

Because the vertical stress is determined as $\sigma'_z = \sigma_z = \gamma h$, where γ is the bulk weight of the rock and *h* is the depth from the surface, it can be finally written for the formation of tensile fractures:

$$c_x < c_y < \frac{1-2\nu}{1-\nu}\gamma h - c_x. \tag{2.1.10}$$

Most of the quantitative studies of the tectonic stress field of the Earth reflect the ongoing contemporary processes, and for them direct or indirect data can be obtained depending of the chosen method of investigation. The presumption for heritability of the tendencies from the past geological times can be applied only for the Quaternary time, under some conditions it could be accepted also the heritability of the Neotectonic tendencies for specific regions. Every reconstruction of the tectonic stress field in older geological formations needs careful analysis of the studied parameters, taking into account the multiform responses of the inverse problem solutions.

When explaining the causes for existing of tectonic stresses and their evaluated rates, it is necessary to realize that the analyses cover only relatively narrow time window and they reflect the contemporary, "the moment" state (in geological sense) of the tectonic stress field. Hence, it is of critical importance to understand what is measured when applying instrumental in situ investigations for stress determination. The terminology used for describing in situ stresses is analyzed by Amadei (1983). More complete and hierarchically structured terminology is presented by Bielenstein and Barron (1971), and it described well the real situation for the near surface geological formations of the Earth's crust (Fig. 2.4).

When analyzing the processes of karst formation and evolution, it is clear that the tectonic factor is the principal controller of the fracturing in the rocks with specific brittleness at given structural position of the rock formations (see Table 2.1). Our experience gives reason to the assumption that the active tectonic

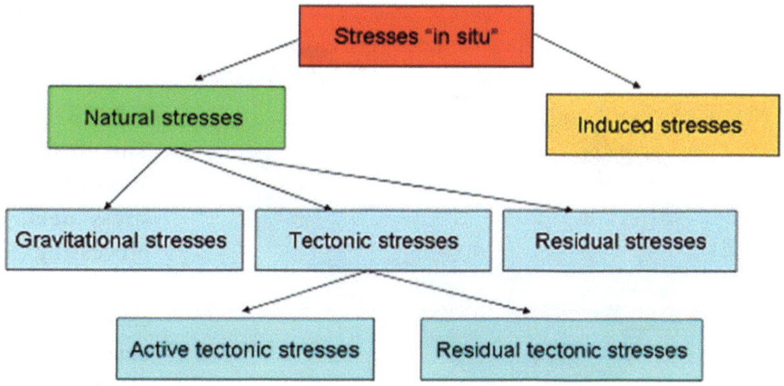

Fig. 2.4 Terminology for the measured in situ tectonic stress fields (following the hierarchy proposed by Bielenstein and Barron 1971)

stresses (Fig. 2.4) during the process of the evolution of the karst systems are the main factor for creating of fissures—tensional fractures or opening of the pre-existing shear joints, in this way facilitating the water circulation and formation of underground galleries with dominant directions.

2.2 Methods of Reconstruction and Analyses of the Tectonic Stress Fields

Among the methods for reconstruction of the spatial direction of the axes of tectonic stress paleo-fields these ones using the fracturing of the rocks in karst areas have shown valuable results. The studies of the striation on tectonic slickensides take a sizable place in the information collected through the years of study on different karst terrains. This type of reconstructions corresponds to the meso-level of the description of the structures, and where the kinematic characteristics of the movements between the rock blocks are used. Here only methods used in the practical works of the authors for analyses of the tectonic stress fields in relationship to the karst systems evolution will be described and discussed.

Another possibility is the analysis of the physical anisotropy when measuring in situ in karst terrains the electrical or the seismic characteristics of the rocks.

The information about the contemporary regional tectonic stress field can be also deduced on the base of the fault-plane solutions from earthquakes at the nearest vicinity of the studied karst terrains.

2.2.1 Striations on Tectonic Slickensides

In these studies, the kinematic characteristics of the movements between rock blocks are used. They could appear at the moment of forming of joints and faults or the orientation of older disruptions could favor the relative movement of adjacent blocks under the conditions of younger tectonic stress field. The orientation of the normal principal tectonic stresses could be reconstructed through meso-structural analysis, using the concept about the relation of movement along shear joints with tectonic stresses (Ramsay 1980; Gamond 1983). The basic working hypothesis, called by the name of Wallace-Both (Angelier 1994) is that if there is a movement and the system "stress $\Longleftrightarrow$ fracturing" is examined independently of environment, then the relative displacement should be parallel to and along the direction of the shearing stress. In Sect. 2.1 the basic relationships have been presented (Griffith's Theory).

The critics of this hypothesis point out as its main defect the fact that, due to the complicated real geometry and the interrelation of faults and joints, a considerable local deviation of the principal normal stresses always appears. The practical analytical results of a great number of conjugated disruption surfaces with movements along them show (Angelier 1994) that amazing constancy is kept and small angle error in the orientation of the shearing stresses τ. The digital modeling (Dupin et al. 1983—cited after Angelier 1994) has shown that the variances depend on the geometry of the intercrossing of the joints, but these variances are very small, within the limits of the error

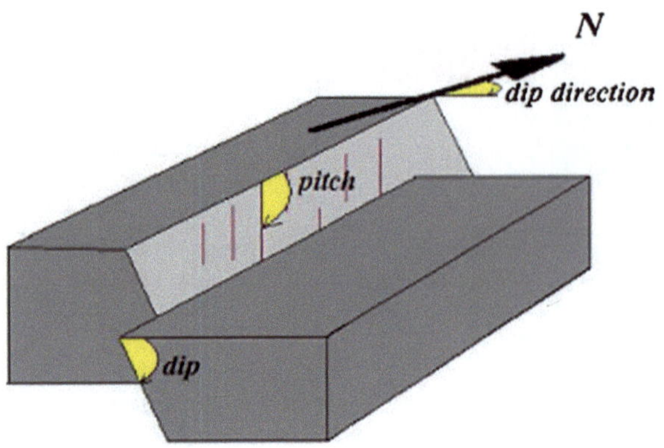

Fig. 2.5 Measured parameters as input datum for the Programme FAULT: dip direction, dip, pitch and sense of movement (*N*—normal, *I*—reverse, *D*—dextral, *S*—sinistral)

of measuring, and practically the movement is always parallel to the medium shearing stress. Thus, the practical usage of Wallace-Both' hypothesis has been confirmed. Numerous researchers created methods and computer programs for reconstructions of tectonic stress directions, using this very hypothesis (good analyses and practical references could be found in Sassi and Carey-Gailhardis 1987; Vergely et al. 1987).

The programme FAULT (Caputo 1989) uses three methods for reconstruction of the principal stress axes by measurements of spatially oriented striations on slickenside and the types of movement on them (Fig. 2.5).

2.2.1.1 Right Dihedron Method

The RDM (after Angelier and Mechler 1977) uses the main hypothesis that the material is preliminary fractured and the size of movement along every single surface is much smaller than the size of the studied rock body. The plane orthogonal to the striation on the disruption surface is defined as a complementary plane. Both planes divide the space around the disruption surface into four rectangular dihedrons (sectors). Every two conjugated dihedrons contain the axes σ_1 and σ_3, which could be presented into a stereographic projection. In a set of measurements, the sectors are defined as zones of action of σ_1 and σ_3 axes.

2.2.1.2 Method of Pressure (*P*) and Tension (*T*) Axes (*P*/*T*)

It is based on the assumption that in homogeneous and isotropic materials, the axis of maximum extension is under 45° toward the disruption surface and the movement direction. Correspondingly, the axis of maximum compression is at 90° toward it.

2.2.1.3 Method of the Smallest Squares

This method complements the procedures for evaluation of the spatial position of the principal axes of stress field, and through it the final result of data processing is formed. The reconstructed principal axes of tectonic stress should be orthogonal toward each other, but they might be dispersed around a center, depending on the data. Exactly, through the method of the smallest squares the position of this center is determined for every principal axis observing the condition for spatial orthogonality between them (Caputo and Caputo 1989).

The main problem that arises at work with striations on slickensides is that the registered movements along various surfaces, even very close (within one outcrop), could not reflect synphase movements. If within the rock volume there is a more than one tectonic impact provoking movement along joints and faults, it is obligatory during the field measurements to find out or at least to suspect the relative temporary sequence of the impacts. The accuracy of the final result is depending from the number of the measured surfaces with striations. It is represented by stereogram (lower hemisphere) with results from RDM and *P*/*T* methods (Fig. 2.6).

2.2.2 Shear Joint Systems

The tectonic fracturing of the rocks gives valuable information about the stress field that caused it and about the orientation of dynamic axes, although it is difficult to extract this information. In spite of the lack of kinematic indicators on most of the joints, they have great potential for reconstruction of the paleostress fields. Especially, this can be used when the joint sets keep consistent spatial orientation on wide areas (Dunne and Hancock 1994). Examples from a platform region were presented and discussed by Shanov (2005). Conjugated joint systems exist in all cases of spatial orientation of the principal axes of particular tectonic stress fields, which should be studied for the reconstruction of the paleo-stresses. In the most simple case, which is observed at preliminary undeformed rocks (as the young sediments are), there is a simple relation between the geometry

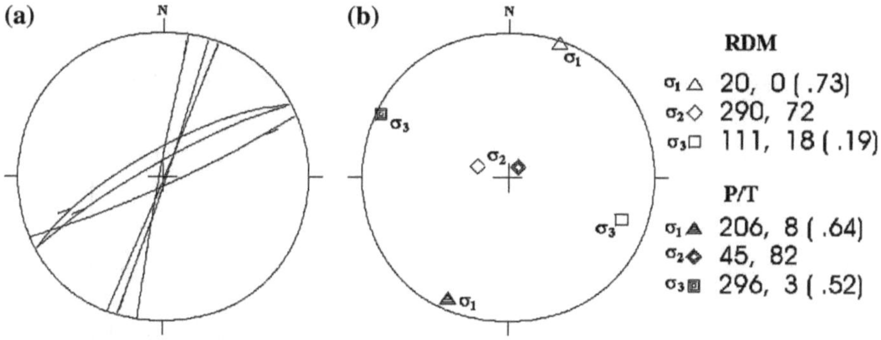

Fig. 2.6 Solution (lower hemisphere) on 6 data from the karst region of Lakatnik (Western Balkan Mountain—Bulgaria). **a** Input data; **b** Reconstruction of the principal stress axes. Close to the symbols for σ_1, σ_2 and σ_3 on the right of the stereonet the azimuth, the dip angles and the Last Square Error (in brackets) are plotted

of the conjugated joints and faults and the stress field, defined with its principal axes σ_1, σ_2, and σ_3 (Angelier 1994). Nikolaev (1977) developed a method for statistical processing and interpretation of joints with good opportunities at various practical structural studies. After analyzing important quantity of joint diagrams, he concluded that the joints' maximums are always distinguished by more or less expressed asymmetry and eccentricity. Accepting that the joints grouped into maximums have been formed at a constant regional tectonic stress field, he concluded that the angle between the conjugated joints changes in a particular way (Fig. 2.7) during the joints' formation.

This change is explained with the fact that the appearance of every new joint causes some change in the stress conditions in its close vicinity. The universal pressure changes and this leads to change in the shear angle θ, at which the next joints originate. Explanation can be easily found if we consider the relationship (2.1.4) from Sect. 2.1. The creation of a new joint leads to decreasing of μ (the coefficient of friction), and consequently the angle θ for every next discontinuity will grow, i.e., the dispersion of the joints will be oriented toward the minimum tectonic stress σ_3.

Similar conclusions were also made by Engelder (1994). According to him, when a pair of shear joints is developed, a torsion of the joint plane appears due to local changes in the stress field, which under particular conditions might lead to the forming of en-echelon segments. As a whole, an asymmetric distribution of frequencies is obtained in the joint sets (Fig. 2.7), which is expressed through the forming of eccentric fields of density on the density diagrams. As

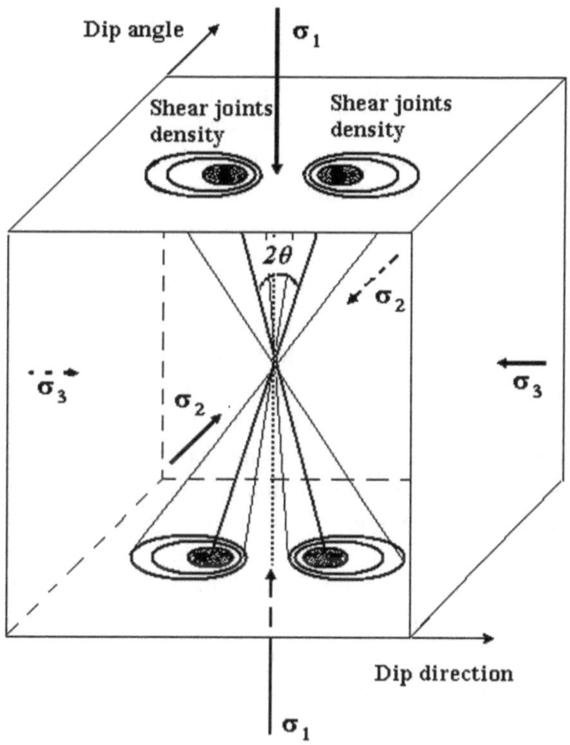

Fig. 2.7 Schematic 3D model of the shear joints' formation and their asymmetric dispersion (following the idea of Nikolaev 1977)

a rule, the extension of iso-areas disperses toward the axis of minimum stress σ_3.

From the Griffith theory (Griffith 1924), it is known that the joints are not distributed in their own plane, but concentrated toward a direction closer to σ_1. This tendency of deviation of the joints' own planes toward the direction of maximum stress is

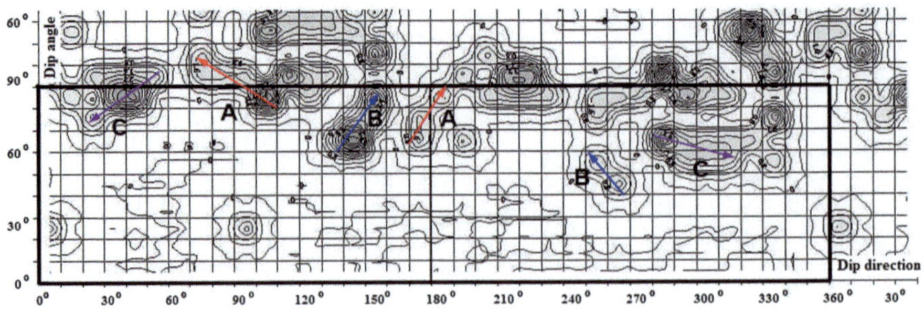

Fig. 2.8 A rectangular diagram of the shear's joints density in the karst area of Cherepish (Western Balkan Mountain—Bulgaria). The sense of dispersion of the pair groups of joints is indicated (*colored arrows*)

shown also experimentally. At conjugated brittle-plastic shear zones, the angle 2θ between them (Fig. 2.7) can vary from 40° to 130° toward σ_1 (Dunne and Hancock 1994). This is especially important for such rock types as dolomites, limestones (the dominant karstified rocks), or alternations of sandstones and more clayey rocks. The usage of joints sets dispersion toward the minimum tectonic stress protects from incorrect determinations of σ_1 and σ_3 axes.

A specific and substantial part of the method is the presentation of measurement data as a rectangular diagram of the joints' density. In a Cartesian coordinate system along one axis, the joints' dip angles are plotted and along the other axis, the azimuth angles of the dip direction are shown. Shanov and Stoyanov (1986) made an analysis and grounded the advantages of Nikolaev's method of presentation. It completely satisfies the conformity requirement regarding the most frequently observed joints: subvertical ones. Regardless of its "matrix" shape, this diagram should be examined also as the transformation of the spherical surface—its projection on unfolded cylindrical surface is tangent to the equator of the sphere.

Finally, the plotting of the centers and the ends of the conjugated joint groups on stereograph projection (Wulf's stereonet) gives the possibility to recognize the principal stress axes σ_1, σ_2, and σ_3. They should be orthogonal and the axis σ_3 is perpendicular to the favorable plane for formation of tensile fractures. Hence, the plane is containing the projection of the axes σ_1 and σ_2. Karst cave networks usually are formed by the water erosion on these planes.

Practically, during field works, it can be recommended to perform between 50 and 100 measurements of joints from the chosen outcrop without giving any priority to some of the existing joint systems. But it is very important to note the coupling of the systems, their spatial superposition, and if possible the relative time sequence. As an example, in Fig. 2.8 is plotted the rectangular diagram of the shear's joints density from the karst area of Cherepish (Western Balkan Mountain—Bulgaria). The interpretation of the results is presented in Fig. 2.9.

Comments

Stress field A: The direction of the maximum compression is NE–SW. It corresponds to the main compression direction during the Middle Laramian tectonic phase. The probability for appearance of open fractures and normal faults striking NE–SW and dipping to SE exists (the gray colored surface).

Stress field B: The maximum compression is directed NNE–SSW. This characteristic corresponds to the advance toward NNE of thrust sheet from South. The formation of new open fractures is of low probability, because their strike is close to the direction of the existing fractures from the previous tectonic impact.

Stress field C: The fracturing is due to the Neotectonic stress field. The deduced surface of normal faulting (in gray color) is corresponding to the geometry of the nearest regional fault. The dominant part of the galleries of the caves of the area follows this

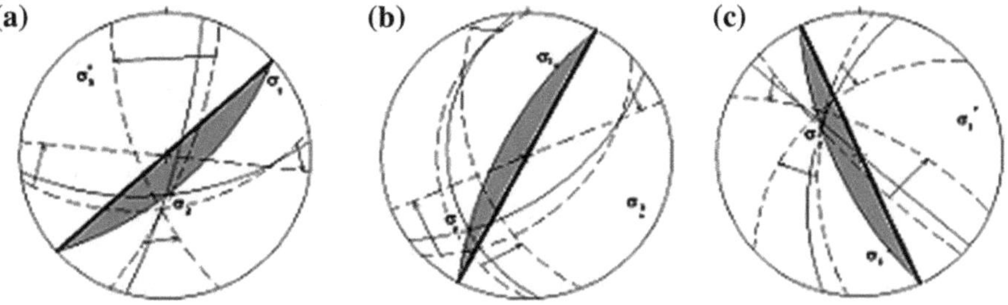

Fig. 2.9 Reconstructions of the tectonic stress fields (lower hemisphere) using the shear joints groups from Fig. 2.8. **a** Paleocene deformations (Middle Laramian tectonic phase). **b** Late Eocene deformations (Late Illyriyan tectonic phase). **c** Neotectonic deformations

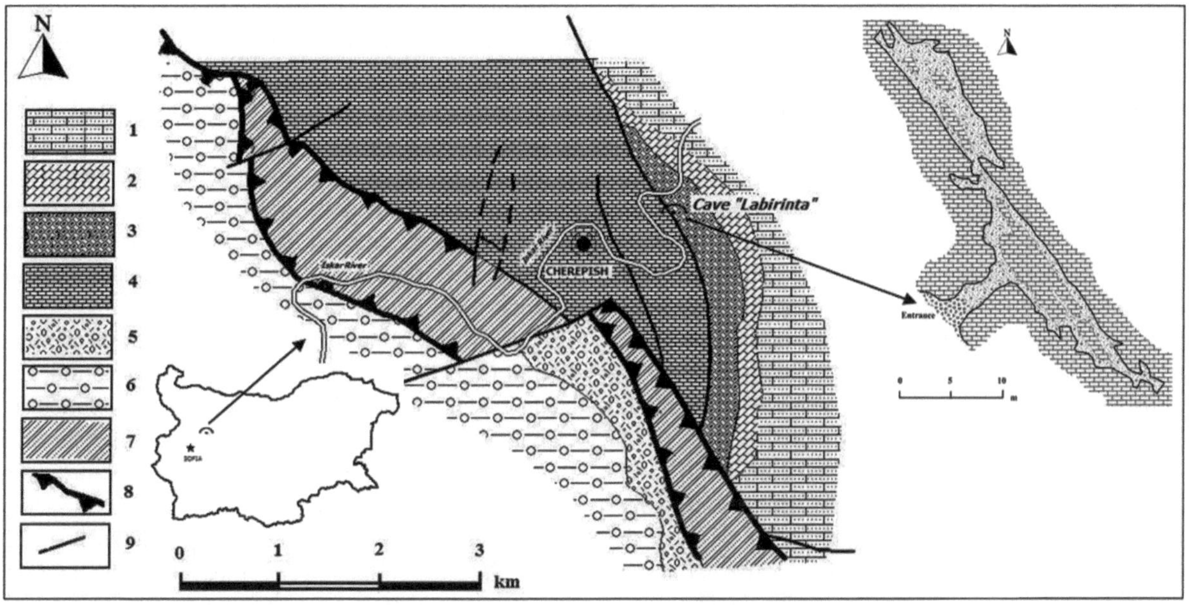

Fig. 2.10 Geological scheme of the karst area of Cherepish (from the Geological map of Bulgaria, Sheet "Vratsa" 1991). The position and the map of the Labirinta Cave are indicated. *Lower Cretaceous*: *1*—Lutibrod formation (oolitic and sandy limestones); *2*—Cherepish formation (massive organogenic limestones); *3*—Mramoren formation (marls with layers of clayey limestones and sandstones); *Jurassic—Lower Cretaceous:* *4*—Slivnitsa formation (massive limestones); *Carboniferous—Permian*: *5*—Vrana formation (breccia-conglomerates, conglomerates); *6*—Buk formation (conglomerates, sandstones, siltstones, argillites); *Cambrian: 7*—Berkovitsa group (metamorphosed argillites, siltstones, marbles and sandstones); *Fault structures: 8*—overthrust; *9*—fault

direction. Especially, the cave "Labirinta" is one of the example of the tectonic control on the formation of preferentially oriented galleries that follow the trend of the open (tensional) fractures deduced from the reconstruction using the shear joints. Here, the impact of the youngest tectonic stress field is evident (Fig. 2.10).

2.2.3 Physical Anisotropy

The physical anisotropy of the rock massifs is the variance of its physical characteristics, depending on the direction of their measurement. The causes for the rock anisotropy can be summarized, as follows (after Uhov 1975):

1. Heterogeneity of the composition and the physical status, caused by the lithological changeability of the rocks in the massif, by the depositional character, by the folding and fracturing in various localities, including big faults, intrusions, fragmentation zones, weathering processes, and water influx.
2. Heterogeneity of the stress condition, caused by the natural stresses in the rock massifs.
3. Large-scale heterogeneity—varies by quality and quantity features of the rocks in studying different volumes from them.

The mineral composition of sedimentary rocks, particularly limestones and dolomites, plays a relatively insignificant role for the anisotropy at meso- and macro-level. The folded structures and the faults impact substantially the physical properties of the rocks, and the resulting rocks anisotropy depends to a great extent on the particular type of deformations and the position of the studied site relatively to these structures. The question about the "scale effect" is also a complex phenomenon and the conformity with it is an obligatory condition for realistic interpretations. For this reason, in geophysical practice, the term "apparent anisotropy" is used, instead of the term "anisotropy" applied mostly for the mineral characteristics (Sheriff 1984).

Two factors are extremely significant at all levels of the studies—the fracturing and the tectonic stresses. As the fracturing is predominantly a result of the impact of the consecutive acts during the geological time of tectonic stresses in the rocks, in fact every study of rock anisotropy in situ could be also presented as a study of the history of tectonic stresses for a particular site. The following working hypotheses should be kept in mind, while the results and analyses presented are obtained from sedimentary terrains predominantly:

1. The fracturing of sedimentary rocks is due to the tectonic stress fields, acting after the rocks lithification.
2. All changes in the direction of principal normal stress axes in space have been reflected in the rock massifs through formation of new joint systems. Our experience has shown that normally more than three sequences of tectonic impacts are difficult to be detected.
3. The anisotropy in the brittle rock varieties (limestones, dolomites, sandstones) is due mainly to the rupture joint system. There is a reason to suggest that after the first-joint system arises, every consecutive by time tectonic stress field will act on already disrupted medium, and will have a weaker deformation effect. The extensional fractures have a definite advantage in anisotropy forming, and their usage at paleotectonic stresses reconstruction is grounded from other authors as well (for example, Letouzey 1986; Caputo and Caputo 1988). Practically, these extensional fractures are presenting the preferential underground routes of the water. The knowledge of the anisotropy in karst areas can be used for determining the dominant direction of the underground karst galleries.
4. In clayey rocks, marls, non-lithified sediments as sands and their varieties, a well-expressed fracturing is not observed, or is missing at all. However, in such kind of materials a compaction along the direction of maximum stress (at regime of compression) or decompaction along the direction of minimum pressure (at regime of extension) is obtained. In both cases, the initial isotropy of the rock material is disturbed.

One of the approaches for anisotropy determination in rocks is to investigate the physical parameters, such as the electrical resistivity. Our experience has provided a number of successful implementation of this type of studies especially on karst terrains.

Some important basic principles are necessary to be presented here for understanding the relationship of the electrical anisotropy measurements and the anisotropic characteristics of the rocks. Theoretical and practical information can be found in many handbooks in different languages. We can suggest the textbook of Sharma (1997) presenting at comprehensible way the modern basis of the geoelectrical survey, including theoretical principles, field procedures and interpretation techniques.

The commonly used method for electrical measurements is to drive an electrical current through the ground and to measure the resulting potential differences on the surface. Electrically better or poorer conducting layers deflect the current and distort the normal potential. Thus, when measuring subsurface variations in electrical resistivity it us possible to detect the anomalous conditions and inhomogeneities within the ground. Different schema of field measurements and related specific methods of interpretations give the possibility to recognize the peculiarities of the electrical properties of the rocks at different depths and to build the corresponding geological model.

The prospection with direct current is widely applied and very useful for determination of the anisotropic characteristics of the rocks. The karstified rocks are, as rule, bad conductors of electricity. Geological processes as faulting, dissolution, shearing, jointing, weathering, and alteration can increase the fluid permeability of the rocks and contribute to the formation of patterns of lower electrical resistivity. In this case, the minerals of the karstified rocks are practically insulators and the electrical conduction is due to the electrolytes (groundwater, wet clayey substance, etc.) filling the pore space and the fissures. If the fractures or the underground karst galleries are without electrically conductive filling (air), these inhomogeneities will have higher (infinite) electrical resistivity relatively to the rock's matrix.

The measurements by direct electrical current can be performed using point or dipole sources. The disposition of the electrodes on the ground determines the geometrical type of the array. The most common disposition of the electrodes is shown in Fig. 2.11.

The aim is to measure the potential difference ΔU between the electrodes M and N when a direct current I is applied between the electrodes A and B. If the resistance between opposite faces of the conducting body with length L and uniform cross-sectional area S is R, the resistivity ρ in homogenous material from single point source is:

$$\rho = \frac{S \cdot R}{L}. \tag{2.2.1}$$

The resistivity SI unit is *ohm meter* (Ωm).

When electrical current I is applied, and the measured potential difference is ΔU, then the relationship with the electrical resistance R is given by Ohm's law:

$$\Delta U = R \cdot I. \tag{2.2.2}$$

Using Eq. (2.2.1) the above relationship can be written, as follows:

$$\rho = \frac{\Delta U}{I} \cdot \frac{S}{L}. \tag{2.2.3}$$

At a distance r, away from the source electrode (A or B), the hemisphere has a surface area $S = 2\pi r^2$, so if $L = r$, and $\Delta U_{MN} = U_M - U_N$:

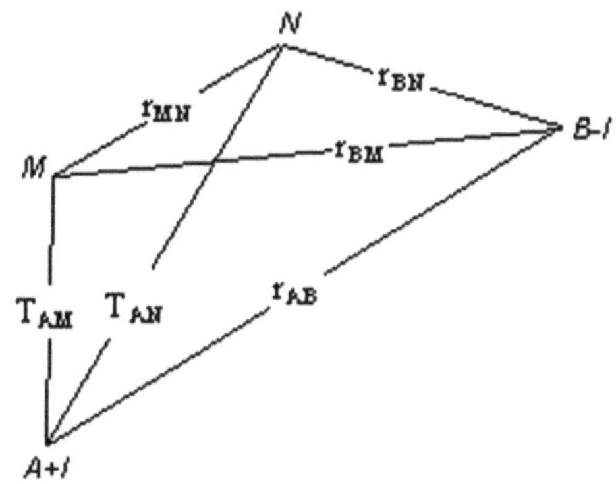

Fig. 2.11 Four electrodes array *AMNB*. *A* and *B* are the current electrodes (C_1 and C_2 in some textbooks), and *M* and *N* (respectively P_1 and P_2) are the electrodes measuring the potential difference

$$\rho = \frac{U_M - U_N}{I_{AB}} \cdot 2\pi r. \tag{2.2.4}$$

The potentials in points M and N are the sums of the potentials from the current electrodes A and B, the electrical current being $+I$ and $-I$, respectively:

$$\begin{aligned} U_M &= U_M^A + U_M^B = \frac{\rho \cdot I}{2\pi} \frac{1}{r_{AM}} - \frac{\rho \cdot I}{2\pi} \frac{1}{r_{BM}} \\ U_N &= U_N^A + U_N^B = \frac{\rho \cdot I}{2\pi} \frac{1}{r_{AN}} - \frac{\rho \cdot I}{2\pi} \frac{1}{r_{BN}}. \end{aligned} \tag{2.2.5}$$

So, the resistivity for uniform homogenous volume in the upper hemisphere is:

$$\rho = \frac{\Delta U_{MN}}{I_{AB}} \frac{2\pi}{\frac{1}{r_{AM}} - \frac{1}{r_{BM}} - \frac{1}{r_{AN}} + \frac{1}{r_{BN}}}. \tag{2.2.6}$$

The relative disposition of the electrodes determine the so-called "coefficient of the array"—k:

$$k = \frac{2\pi}{\frac{1}{r_{AM}} - \frac{1}{r_{AN}} - \frac{1}{r_{BM}} + \frac{1}{r_{BN}}}. \tag{2.2.7}$$

Practically, at meso- and macro-level, the rocks are neither uniform, nor homogenous. At given position between the electrodes, the electrical current passes trough composite rock material with imposed secondary brittle or ductile deformations. This fact argues the use of the term "apparent resistivity" ρ_a for the measured in situ resistivity. The final equitation becomes:

$$\rho_a = k \frac{\Delta U}{I}. \tag{2.2.8}$$

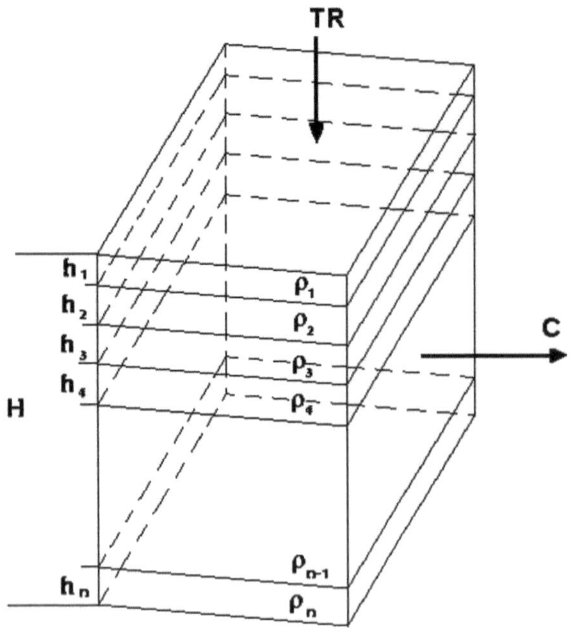

Fig. 2.12 Model of a sequence of horizontal layers

The apparent resistivity is a function of the specific electrical resistivity of the rocks, the thickness and the depth of the layers and other rock inhomogeneities as fractures and the type of their filling substance. Thus, the electrical field should be not symmetric in the space. Its configuration will depend on the superposition of the electrical conductivity of the studied "package" of layers and structures. The same is valid for the electrical resistivity.

Let us consider the simplest case, representing a package of horizontal layers with electrical resistivity of every layer ρ_1, ρ_2, ρ_3, … ρ_N (Fig. 2.12).

In this model $L = 1$. The thickness of the layers is h_1, h_2, h_3, … h_N. From Eq. (2.2.1) is seen that the resistance R is:

$$R = \frac{\rho \cdot L}{S}. \tag{2.2.9}$$

The parallel conductivity C_i when the electrical current is applied on the transversal walls of every layer i is:

$$C_i = \frac{1}{R_i} = \frac{h_i}{\rho_i}. \tag{2.2.10}$$

The transversal resistance TR_i when the electrical current is applied from the top and the bottom of the layers' package is:

$$TR_i = R_i = \rho_i h_i. \tag{2.2.11}$$

From Eqs. (2.2.10) and (2.2.11) it can be determined that:

$$\begin{aligned} h_i &= \sqrt{C_i TR_i} \\ \rho_i &= \sqrt{\frac{TR_i}{C_i}}. \end{aligned} \tag{2.2.12}$$

For the total thickness $H = \sum_{i=1}^{N} h_i$ the total parallel conductivity C and the total transversal resistance TR are:

$$\begin{aligned} C &= \sum_{i=1}^{N} \frac{h_i}{\rho_i} \\ TR &= \sum_{i=1}^{N} h_i \rho_i. \end{aligned} \tag{2.2.13}$$

The average parallel resistivity ρ_l and the average transversal resistivity ρ_t will be:

$$\begin{aligned} \rho_l &= \frac{H}{C} \\ \rho_t &= \frac{TR}{H}. \end{aligned} \tag{2.2.14}$$

It is proven that always the transversal resistivity ρ_t is higher than the parallel resistivity ρ_l (Hmelevskoy 1984).

Let us now consider a more complicated and close to the practice case, when the layers are dipping from the surface at angle α and the measurement of the electrical resistivity is executed using the symmetric four electrodes array Schlumberger (Fig. 2.13). The array can be oriented randomly on the surface, crossing the parallel layers at angle β.

Without entering in the theoretical details, it can be shown (Yakubovskiy 1973) that:

$$\rho_a = \frac{\rho_l \sqrt{\rho_t}}{\sqrt{\rho_l(\cos^2\beta + \sin^2\beta \cdot \cos^2\alpha) + \rho_t \sin^2\beta \cdot \sin^2\alpha}}. \tag{2.2.15}$$

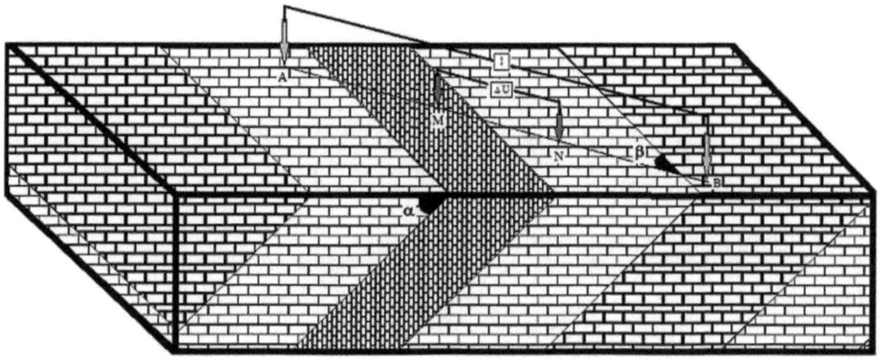

Fig. 2.13 Model of field measurement of the electrical resistivity of layered rocks

Two extreme case exist—when the array is parallel to the layering (or to the extension of the inhomogeneities), i.e., $\beta = 0$, and when the array is perpendicular to the layering ($\beta = \pi/2$). For the first case the maximum apparent resistivity will be measured, because:

$$\rho_a(0) = \sqrt{\rho_l.\rho_t} = \rho_{\max}. \quad (2.2.16)$$

For the second case the apparent resistivity is lower

$$\rho_a(\pi/2) = \frac{\sqrt{\rho_l \rho_t}}{\sqrt{1 + (\lambda^2 - 1)\sin^2\alpha}} = \frac{\rho_{\max}}{\sqrt{1 + (\lambda^2 - 1)\sin^2\alpha}}. \quad (2.2.17)$$

Here λ is the coefficient of macro-anisotropy, because it characterizes the integral anisotropy of the studied rock volume and it is $\lambda > 1$:

$$\lambda = \sqrt{\frac{\rho_t}{\rho_l}}. \quad (2.2.18)$$

It is evident, that $\rho_a(0) > \rho_a(\pi/2)$. The measured apparent resistivity parallel to the elongation of the inhomogeneities is higher then perpendicularly to them. So, this is the "paradox of the anisotropy," because in reality the transversal resistivity ρ_t is higher than the parallel resistivity ρ_l.

In practice, using arrays with different azimuths around a given point and with different distance between the electrodes *A* and *B*, it is possible to obtain the electrical anisotropy ellipse for different depths (Azimuthal Vertical Electrical Sounding). For different depths (different stratigraphic layers), the ellipse of the electrical anisotropy can be obtained, which is closely related to the rock bedding and fracturing through the "paradox of anisotropy." In most of the practical cases, the longest axis of the electrical anisotropy in horizontal plane is coinciding with the dominant direction of the tensile fractures, normally controlling the general orientation of the karst galleries.

As example, a study in karst area in Bulgaria is presented here.

2.2.3.1 Rumiantsevo Region (Central North Bulgaria)

The investigation aimed to study the space position of 3 m high karst cavity crossed by an exploration well at a depth of 19.6 m below the surface. The studied area is located near the village of Rumiantsevo (North Bulgaria). This is a potential area for a future mining of limestones from Kailaka Formation. It is built of white, massif, and organic limestones with various biological inclusions. The age is considered to be Upper Maestrichtian. Limestones are easily karstified near the fractured zones and faults. The region of nearby Karlukovo village is known with its numerous surface negative karst forms and many caves (Fig. 2.14).

Above-mentioned modification of Vertical Electrical Sounding (VES) method was used to applying an azimuth scheme on the site where Well No.10 crossed an underground cave. An important data on the electric anisotropy of the rock at different depths and prevailing direction of the fractures could be obtained by using this scheme of measurements. VES 1 is coinciding with the wellhead. Knowing the real depth of the layers and the crossed cavity, the survey at this site was as a "parametric" VES, because the information from the well was used for calibration of the rock resistivity (Fig. 2.15). When fixing the known depths, it was possible to evaluate the real pseudo-resistivities of the cross-section. Nevertheless, the automatic interpretation of the curve by the

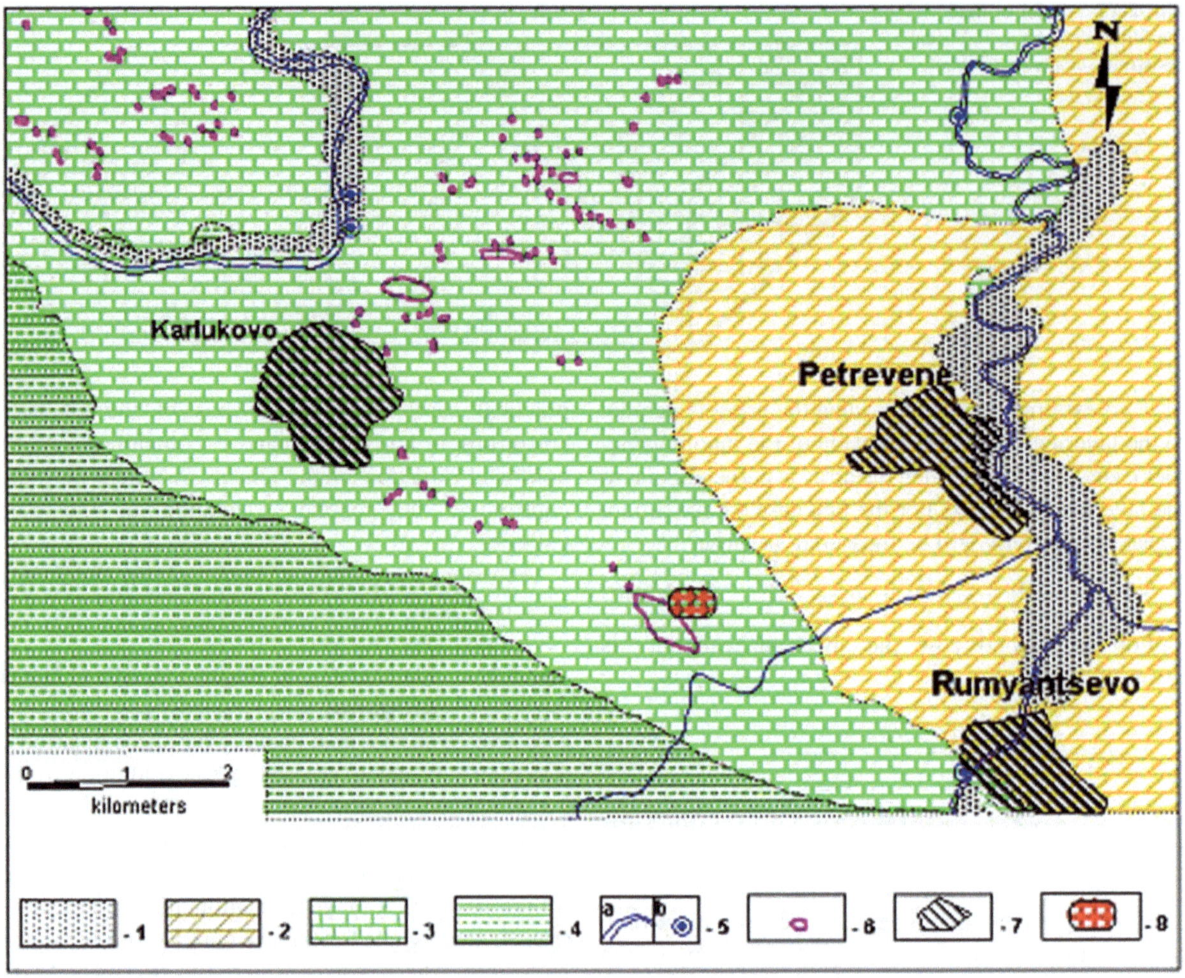

Fig. 2.14 Geological map of the studied area near the village of Rumyantsevo (according to Tsankov 1991): *1*. Alluvial deposits—sands, gravels, clays; *2*. Paleogene marls, sandstones, siltstones, clays; *3*. Upper Maestrichtian limestone; *4*. Lower Cretaceous sandstones, marls, siltstones; *5*. **a** rivers; **b** karstic springs; *6*. dolines; *7*. villages; *8*. studied area

program IPI2WIN gave the position of the cave between 22.5 and 29.9 m from the surface. The only point on the electrical curve showing the presence of the cave is at $AB/2 = 25$ m. This length of the electrical current supplying array was used for the Method of Dipole Profiling.

Two types of limestones (according to their electric properties) are present in the geological cross-section of the area. Limestones of comparatively high electric resistivity (1,000 Ωm) are defined near the surface, particularly in the northeastern part of the area. They could be identified with the third horizon of Kailaka Formation. Limestones of lower resistivity (400–600 Ωm) are disposed below them with a top at about 11–15 m depth from the surface. They could be identified with the second horizon of Kailaka Formation. Karst forms are found only in the horizon of lower resistivity. No available data exist for presence of such forms in the overlying limestones of higher resistivity. Electric resistivity of karst cavities is high (practically tends to infinity). According to VES curves, ρ_a is in the range of 4,000–5,000 Ωm.

Measurements at 4 azimuths (N 0, 45, 90, and 135°) were carried out near the wellhead. The result from the array $AB/2 = 25$ m is very important. It gave information from the depth of the karst cavity crossed by the well (Fig. 2.16). Abrupt changes of electric parameters (coefficient of anisotropy, eccentricity, direction of the long axis of the anisotropy ellipse and electrical resistivity) of the rocks were

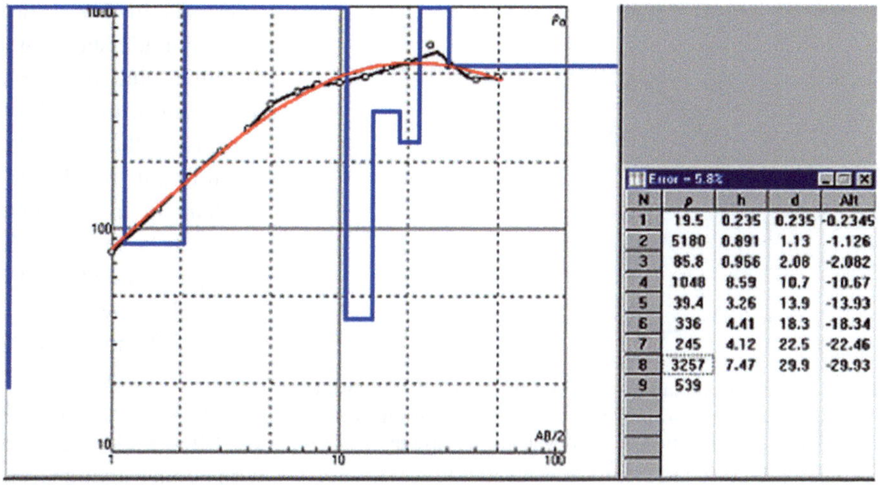

Fig. 2.15 Interpretation of VES 1 curve (Software IP2WIN). The thickness of the layers is according to the well log. The error of the approximation is 5.8 %

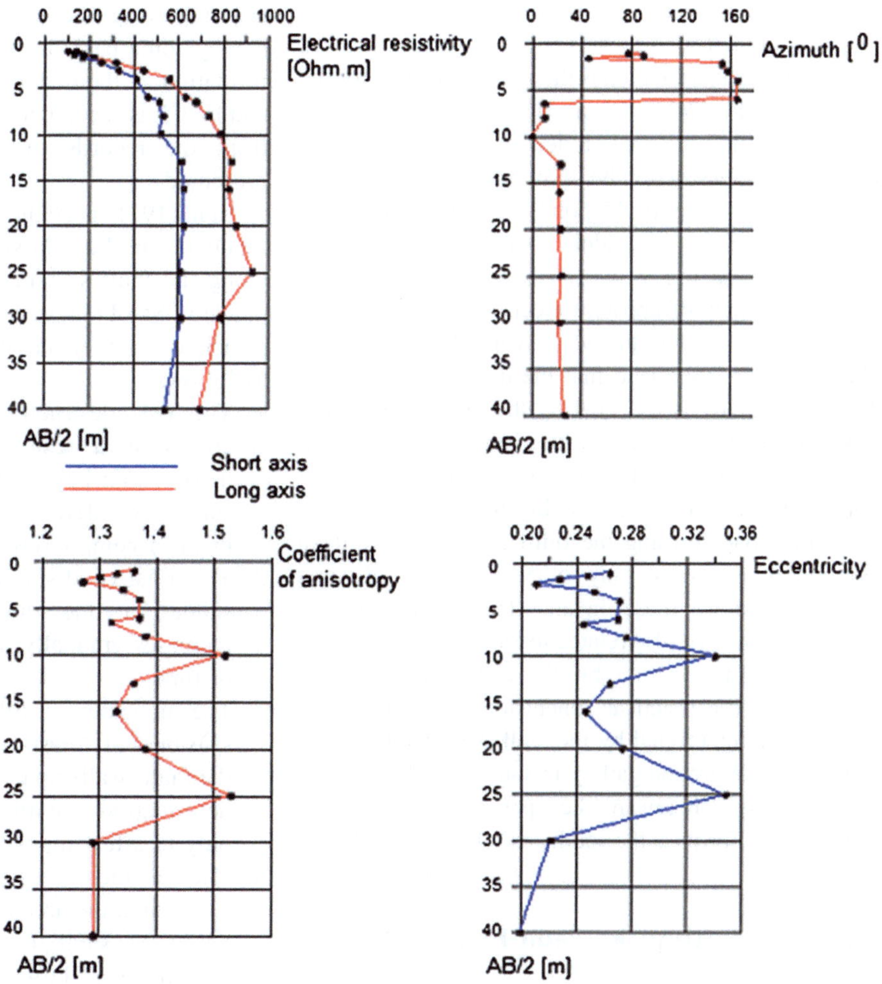

Fig. 2.16 Graphic characteristics of changing electric anisotropy in limestone layers with depth

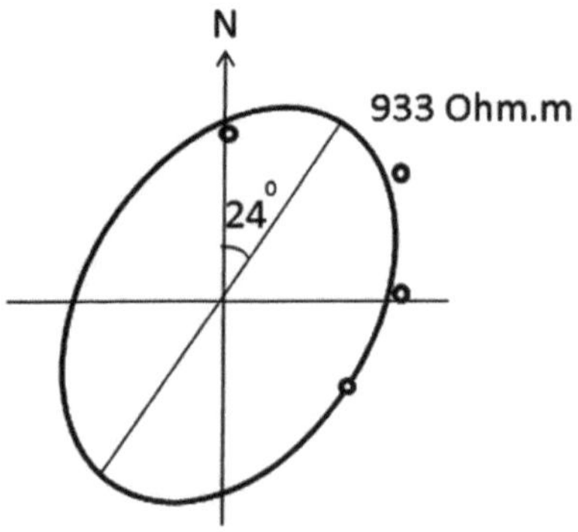

Fig. 2.17 Ellipse of anisotropy of the apparent resistivity values measured by array $AB/2 = 25$ m at 4 azimuths (Rumyantsevo, North Bulgaria). Direction of the long axis of anisotropy—N 24°; Coefficient of anisotropy—1.29; Eccentricity—0.348

detected at depths corresponding to arrays $AB/2 = 10$ m and $AB/2 = 25$ m.

Lithological alternation is the possible reason for the shallow level ($AB/2 = 10$ m), while for greater depth ($AB/2 = 25$ m), this could reflect only the presence of karst cavity elongated in NE–SW direction (Fig. 2.17).

Dipole profiling in 65 points located in the area using the scheme of Middle Gradients was applied with 50 m spacing between the current electrodes *A* and *B*. A map of apparent electric resistivity of the southern part of the studied area has been compiled (Fig. 2.18). It is evident, that the discovered by Well 10 karstic cavity is reflected by the elongated toward SW electrical anomaly with values higher then 850 Ωm. But the very high resistivities in the northern part of the studied area, probably related to near surface limestone forming the third horizon of Kailaka Formation, can mask any other existence of karst structures below these layers. The high electrical resistivity could be also related to karst structures in the upper part of the section, but this can be verified only by drilling. The practical conclusion is that the karst cavity zone intersected by the well is probably developing as underground gallery in southwest direction. It drains the system toward the open surface karst and underground forms at the southwest of the studied area.

2.2.4 Earthquake Fault-Plane Solutions

The knowledge of the processes inside the earthquakes foci at first look is not evidently related to the processes of karst formation and evolution. But a broader understanding of the contemporary Earth's dynamics and its direct impact on the faulting and fracturing of the rock volumes leads to the inevitable conclusion that these two phenomena are final results at different levels of the tectonic stresses. The mechanism in the focus of every earthquake is an informative source today about the regional (strong earthquakes) and local (weaker earthquakes) tectonic stress characteristics. It has been demonstrated in Sect. 2.2.2 how the tectonic stresses control the principal tendencies of the formation of the underground karst systems of galleries. Furthermore, if the karstified rocks are situated inside a tectonic province with contemporary expressed activity, the earthquakes could be the factor for secondary moment deformations of the speleosediments. This phenomena and its significance is discussed in Chap. 3. That is why some basic information on the mechanism of formation of the dislocation in the earthquake focus, and how it is recognized by "the fault plane solution" from the records of the seismic waves will be presented.

Reid (1911) presented his theory about the "elastic rebound" in his classic study on the dislocations along the fault system San Andreas after the earthquake in San Francisco in 1906. Reid formulated five principal statements in the theory (after Stacey 1972), as follows:

1. The rocks destruction causing tectonic earthquake comes as a result of accumulation of elastic deformation beyond the limit that rock can withstand. The deformation arises at relative movement of neighboring blocks in the Earth's crust.
2. The relative displacement of the blocks is not sudden at the moment of disruption, but it increases gradually for a longer or shorter periods of time.
3. The movement at the moment of earthquake consists only of "elastic rebound"—a sharp moving of the fault' walls up to the position, where the elastic deformations are missing. This movement is visual only to a few miles far from the fault.
4. The seismic waves originate out of the fault surface. First, the area of the surface where they arise out is very small, but after that it increases fast and becomes very large, but the velocity of its propagation does not exceed the shear wave's velocity of the rocks.

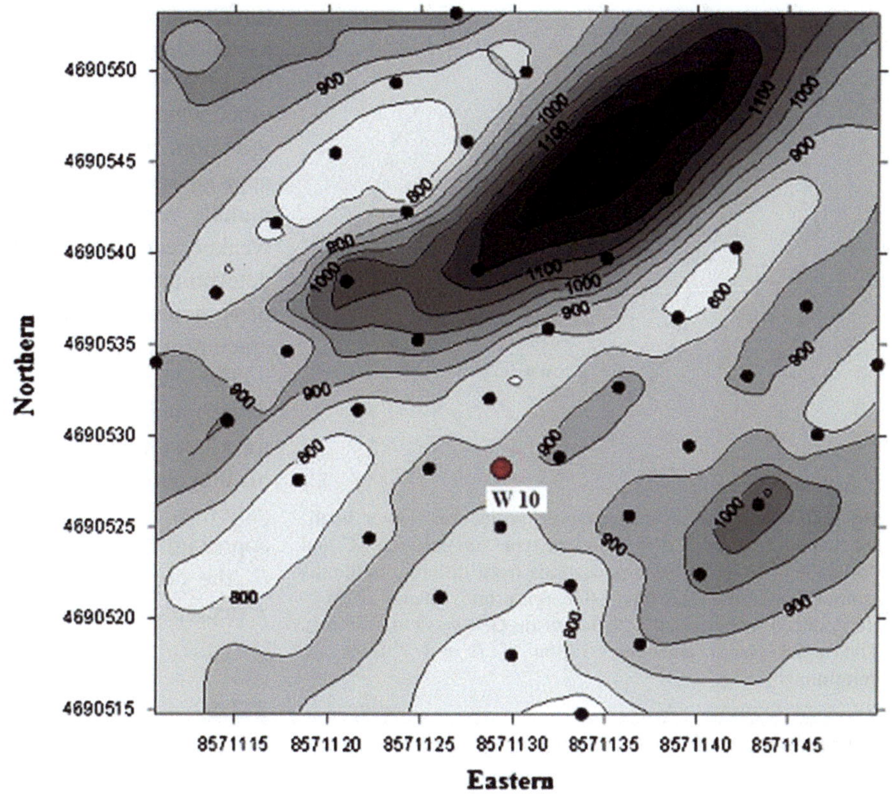

Fig. 2.18 Detailed map of the apparent electric resistivity anomaly zone around *Well 10*. The *dots* are the points of measuring of the apparent electrical resistivity

5. The energy released during the earthquake, immediately before the earthquake, is the energy of the elastic deformation of the rocks.

This theory also initiates the development of the Dislocation theory of earthquakes. Without any details upon the theory (very well shown in the works of Stacey 1972; Cox and Hart 1989, and others), it is important to note that this theory created the basis for explanation of the character of the emitted seismic waves, depending on the type of movement along the dislocation. The main idea is to present the final dislocation in the earthquake focus as an equivalent of a pair of stresses with or without moment. There is no united opinion about which of both cases describes more precisely the processes in the focal zone. In any case, the pair of stresses with moment describes better the arising of medium disruption, moving along it, and creation of waves of "contraction" and "dilatancy" (Fig. 2.19), which are registered with different polarity of their first breaks on the seismograms.

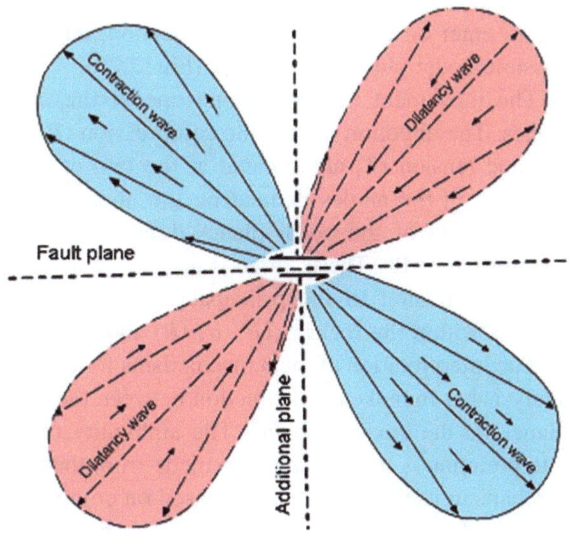

Fig. 2.19 Formation of waves of "contraction" and "dilatancy" in the earthquake focus (following Drumia and Shebalin 1985)

Usually, the study of the seismic source is through analysis of the seismic waves' records on seismic stations, outlying from the focal zone at a distance bigger than the size of the focus. It means, loss of a considerable quantity of information due to fading of the high frequencies with the distance. Nevertheless,

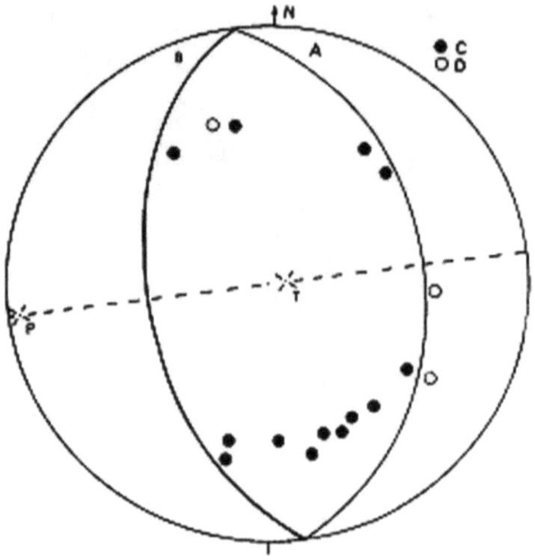

Fig. 2.20 Example of stereographic projection (lower hemisphere) with recorded first brakes of type "contraction" C and "dilatancy" D by the seismic stations from different azimuths around the earthquake focus (Georgiev and Shanov 1991—Reproduced by permission of *Bulgarian Geophysical Journal*). The nodal planes and the position of *P* and *T* axes are determined

the parameters of the focus and the processes in it are calculated, sufficiently, for the tectonic studies accuracy through these records. The regions of positive and negative first breaks (first arrival) of longitudinal waves enter on circle in Wolf' Projection, and both possible nodal planes are defined (Fig. 2.20).

The movement occurred along one of the nodal planes. The direction of tectonic compression (*P*) and tectonic tension (*T*) are situated at the two opposite bisectors of the angles defined by the nodal planes. The crossing of the nodal planes defines the intermediate axis B. These three axes are practically indirectly related to the principal axes of tectonic stresses within the rocks $\sigma_1(P)$, σ_2 (B) and $\sigma_3(T)$.

The interpretation of focal mechanisms for tectonic analysis has to make the distinction between the fault-plane and the auxiliary plane. The ambiguity of this interpretation is that it cannot be made from the focal mechanism itself and it must be based on comparison with the local geological structures. The main problem is connected with the synonymous determination of the nodal plane, along which the movement is performed, and it could be realized through additional tectonic and geophysical information only. The other uncertainty arises in attempting to relate the *P* and *T* axes with σ_1 and σ_3 axes of the tectonic stress field. Various inversion schemes have been developed (Gephard and Forsyth 1984; Michael 1987). The assumption of these methods is that the orientations of the principal stress axes are constant within the volume, which is being analyzed. Such dynamic interpretation of focal mechanism data is not always appropriate (Scholz 2002). For some areas, limited number of earthquakes has fault-plane solutions, for weaker earthquakes ($M < 4$) it could be expected determination of the focal mechanism only in the case of sufficient records from seismic stations around the epicentral area.

For the karst areas, the nearest earthquakes with determined fault-plane solutions can give useful information about the tendency of spatial orientation of the active (with water stream) galleries. These galleries, by their definition, have to be oriented approximately at along the same direction as the axis *P* (the compression) or perpendicularly to the axes *T* (tectonic extension).

2.2.5 Time Sequence of the Reconstructed Stress Fields

When working in karst areas, any collected information concerning the reconstructions of the stress fields can be of importance for ranging the tectonic impacts consecutively in the geological time scale, and to make the conclusions for the relationship between tectonics and karst evolution. Normally, a number of paleotectonic stress fields can be reconstructed. The first of evident bench marks for ranging them in the time is the age of the concerned karstified rocks. The first impact forming fractures had been obligatory manifested after the formation and lithification of the rocks, not earlier. The next steps are to find reconstructions of the stress field in the area in younger sediments than these containing the karst system. These sediments can provide information about the stress fields after their formation, but the deformations obligatory have impacted the older rock sequences. The last step is to compare the reconstructions of the stress field with the fault-plane solutions from nearest earthquakes, if available. These solutions give immediate information closely corresponding to today stress conditions and they can help the correct interpretation. The appropriate arrangement in time of the deformations can be successful only when using all available geological, geomorphological, hydrogeological, geophysical, and geodetic

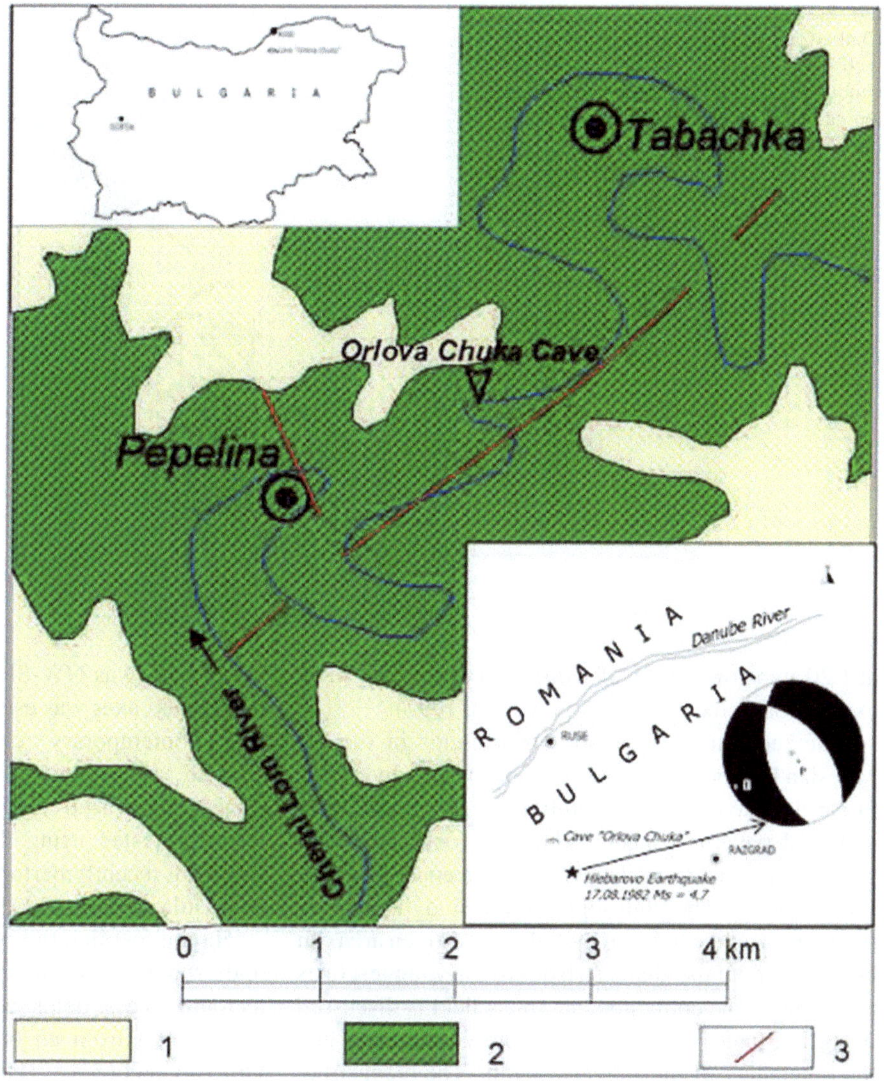

Fig. 2.21 Regional scheme, nearest earthquake fault-plane solution and geological background of the Orlova Chuka Cave area (North Bulgaria). The fault-plane solution for Hlebarovo Earthquake ($M_s = 4.7$) is based on the data published by Shanov et al. (1988). The geological map is based on Sheet "Byala" from the Geological Map of Bulgaria in scale 1:100 000 (Filipov 1992) *1*—Loess cover; *2*—Aptian limestones (Lower cretaceous); *3*—fault

information. The maps of the known karst galleries are necessary to make the most possible realistic conclusions for the features of the karst formation and evolution in the area concerned.

An example from the Moesian Platform (Northern Bulgaria) is presented below.

Orlova Chuka Cave, North Bulgaria

The cave *Orlova Chuka* (means Eagle rocky peak) is located on the left bank of Cherni Lom River (Fig. 2.21). The cave is a complex system of galleries, caves, and large and small halls. It is the second in Bulgaria by the total length of its galleries—13,437 m. The cave is formed in Cretaceous (Aptian-Urgonian) limestones and calcareous sandstones. According to Radulov (2002), the morphological features of the cave galleries suggest a phreatic origin. The cave system lies just below an old river terrace at elevation about 60 m above the present floodplain of Cherny Lom River, tributary of Danube River. Terrace at 60 m corresponds to an erosion surface formed in the southern margin of the so-called Dacian Basin during the Pliocene and Early Pleistocene. Erosional surface is cut 2.59 Ma BP, and the alluvium deposits over it extend to the beginning of loess formation in NE Bulgaria. The main phreatic stage of cave formation should be assigned to the same time span. Brunhes/Matuyama geomagnetic boundary (0.78 Ma BP) has been found

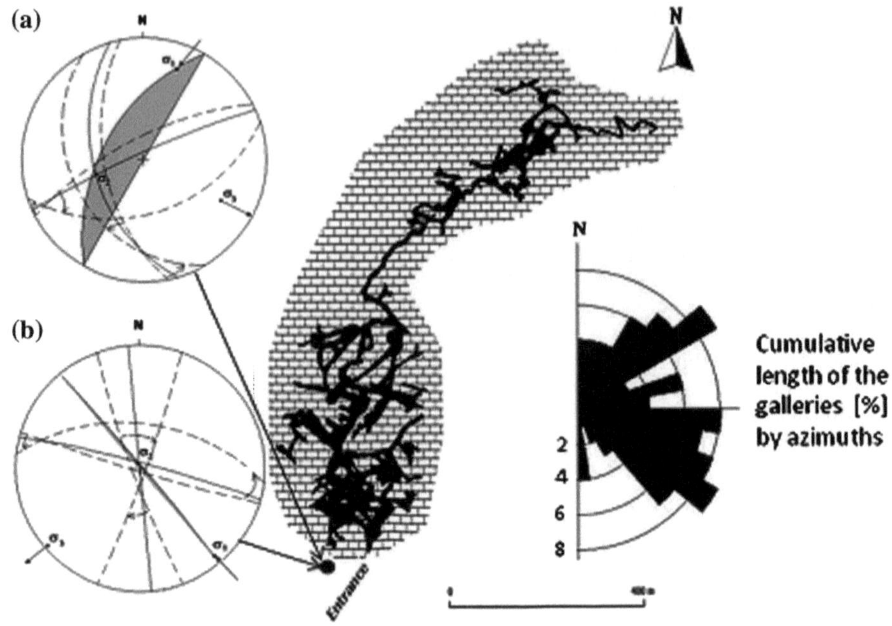

Fig. 2.22 Map of the Orlova Chuka Cave (North Bulgaria) with rose-diagram of the cumulative length (in %) of the galleries, compared to the directions of the open (tensile) fractures (in *gray* on diagrams *A* and *B*) generated by the tectonic stress fields in the Aptian limestones

at 1.20 m below the top of sedimentary sequence in an excavation in the cave (Evlogiev et al. 1997).

The reconstruction of the stress fields in the SE Moesian Platform in Bulgaria has been made by means of earthquake fault-plane solutions, tectonic fractures, and fold patterns (Shanov 2005) in rocks restricted stratigraphically from the Early Cretaceous up to the Late Pliocene. The philosophy of the study is that the recognition of the older stress fields for a given area can be successful if the characteristics of the youngest ones are known. The contemporary stress field is discussed using the fault-plane solutions from earthquakes and the kinematics of the activated faults. The reconstruction of the Post-Pliocene paleo-stress field was made by studies of conjugate shear joints systems in Upper Pliocene limestones, covering the area east of the location of the cave and containing traces from only one tectonic deformation (Post-Pliocene).

Measurements of elements of tectonic fractures were also performed in situ on more than 60 outcrops of rocks of Aptian age. A more complete study of the tectonic meso- and micro-structures has been performed on the Sarmatian sediments in the SE Moesian Platform. This study includes a description of the discovered folds, brittle tectonic analysis, and reconstruction of the Post-Sarmatian paleo-stress field. As a result, it was deduced that the compression after the Early Cretaceous period is NE-SW directed. The direction of compression since Sarmatian to Early Quaternary was NW-SE. A clockwise rotation of the main stress axes was established for a number of sites. The contemporary contraction is directed also NW-SE, according to the fault-plane solutions determined for crustal earthquakes in the region. This result is tested using the data from the GPS measurement recently performed in this part of the Balkan Peninsula (Shanov 2005).

Having in mind this regional analysis, it is easier to understand the results from the reconstruction of the tectonic stress fields using measurements of joints (106 joints) from an outcrop at the entrance of the Orlova Chuka Cave (Fig. 2.22).

At the first stage of its formation, the Post-Cretaceous fracturing (the age is not evaluated but the deformation happened before the end of Sarmatian time) probably had some role to predestine the NE–SW trend of the open fractures, able to form karstic galleries (solution *A* in Fig. 2.22). In any case the longest gallery in the cave is parallel to the regional fault striking NE–SW.

The second impact *B* was due to the more recent deformations, probably related to the Post-Sarmatian processes. The Post-Sarmatian (Late Neogene?) deformations marked a total change of the preferential direction of cave formation due to the interchange of the orientations of σ_1 and σ_3 tectonic stress axes. Since this time, because the tectonic regime as general tendencies has not changed till present, the main tectonic stress field axes keep their general orientation and the consecutive dominant NW-SE direction of the largest galleries in the cave. The main phreatic stage of cave formation and especially the formation of the large halls can be assigned to this period. The today stress field has not impacted the galleries trend, because the cave has been not active at least during the last 0.7 Ma.

The comparison of the reconstructed tectonic stress fields with the dominants directions of the galleries of the Orlova Chuka Cave confirms totally the tectonic control of the cave formation (Fig. 2.22).

2.3 Tectonic Stress Control on Karst Systems: Case Studies

The presented case studies are from different countries and different tectonic context had impacted the formation of the karst systems. Nevertheless, the tectonic control is always present and the contemporary processes of karst evolution show clear relationship with the youngest fracturing, i.e., with the recent tectonic stress field characteristics.

2.3.1 Albania: Tectonic Factor for Karst Formation in the Albanian Dinarides

2.3.1.1 Short Geological and Tectonic Characteristics

The structural complex characterizing the exceptionally complex geological structure of the not so large Albanian territory is named by Albanian geologists Albanides (Biçoku et al. 1978). Both basic mega-structural units of the Albanides—North Albanides and South Albanides (related with the Helenides) are separated by the Shkodra - Pejio transversal zone representing an old paleogeographic province. The tectonic domains display different deformation styles. The External domain, where the presented studies were performed, including Paleogene and younger units of the Krasta–Çukali and other tectonic zones to the west, has been affected by compression, manifested in west-directed folds, thrust faults and oblique strike-slip fault systems (Dilek and Koçiu 2004). The karst of Albanian Dinarides (Albanian Alps zone according to some authors) is considered as continuation of the High Karst of Dinarides (Fig. 2.23).

The lower part of the geological section is represented by terrigenous sediments of Permian to Lower Triassic age. They are covered by carbonate sediments. Their accumulation began during the Middle Triassic. Bauxites were formed in the same time. A continuous sedimentation of neritic limestone followed to Maastrichtian time. Jurassic pelagic limestones with silicitic layers occur in the eastern part of the Albanian Dinarides. This part (called also Valbone Subzone) is considered as an independent transitional element of uplifted Vermosh flysh, which covers a period from Maastrichtian to Eocene.

At the end of the Eocene, the zone of the Albanian Dinarides was deformed and thrusted to the south over the zone Krasta-Çukali. From structural point of view, the Albanian Dinarides are a monocline dipping northwest. The large Valbone anticline is located to the northeast and the large Malesia-Madhe syncline is situated to the northwest.

The region in which the expeditions of the Bulgarian Speleological Federation worked during four consecutive years (beginning from 1990) is a part of the North Albanides. According to the age of the more intense folding, the region belongs to the areas deformed at the end of the Eocene. More detailed structural and tectonic investigations in 1994 have been carried out on the territory of the SE limb of Malesia-Madhe syncline, and especially near its axial part. The geology of this territory is represented mainly by carbonate complexes of Lower Jurassic up to Lower Cretaceous (Barremian-Aptian) age. The results of the structural-geological studies were published by Shanov (1999). Here, the tectonic control on the karst formation in the studied area is discussed following this publication.

Different-order faults determine the formation of the tectonic block structure of the limestone massif. They are long-term acting faults with quite explicitly manifested neotectonic activity affecting the relief and the karstic processes in the region. The recent activity of some of the faults is manifested by concentration of earthquake foci along them (Fig. 2.24). The fault Rapsh- Boçani is quite active during the recent tectonic stage (No. 4 in Figs. 2.24 and 2.25).

It is clear, that energy accumulation occurs along this fault, periodically being released by earthquakes

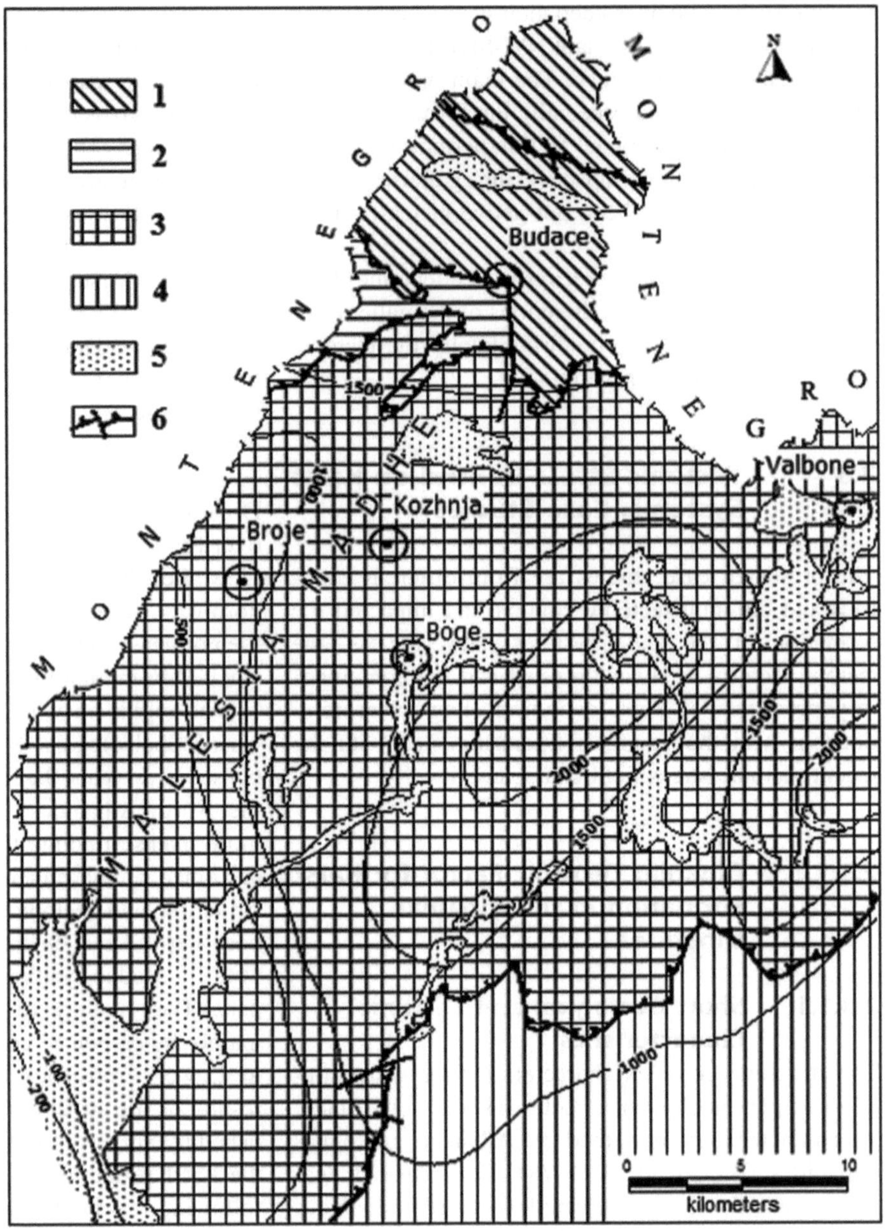

Fig. 2.23 Regional tectonic scheme of North Albania (after Biçoku and Aliaj 1973a, b and the Geological Map of Albania in scale 1:200 000—Harta Geologjike 1981). Isolines of the maximum uplift in metres during the Neotectonic stage are shown (Shanov 1999—Reproduced by permission of *Geologica Balcanica*) *1*—Zone Gashi, uplifted after Early Paleogene; *2*—Zone of the High Karst, uplifted after the end of early Paleogene; *3*—Zone of Albanian Dinarides (Albanian Alps), uplifted during the first half of the Paleogene; *4*—Zone Kras-Cukali, uplifted during the second half of the Paleogene; *5*—Neotectonic superimposed depressions; *6*—thrust front lines

in the zones of its crossings with other regional faults (e.g., with fault No. 1—Pjetroshan- Velicikut-Kozhnje, or with fault No. 3—Böge-Budace). Following the most general considerations reflected on the geological map of the region, it could be assumed that, in the contemporary tectonic stage, there is an uplift of the northern adjacent to the fault block and subsidence of the southern one. This is a continuation of the tendency of neotectonic development of the region, which has predestined the contemporary relief. The neotectonic stage is marked by intense uplift total amplitude in some parts more than 2,000 m.

2.3.1.2 Geological and Hydrogeological Conditions of Karst Formation

During the whole Mesozoic after Early Triassic time, the depositional settings, existing almost continuously up to the beginning of the Maastrichtian, resulted in the formation of a thick limestone complex. The

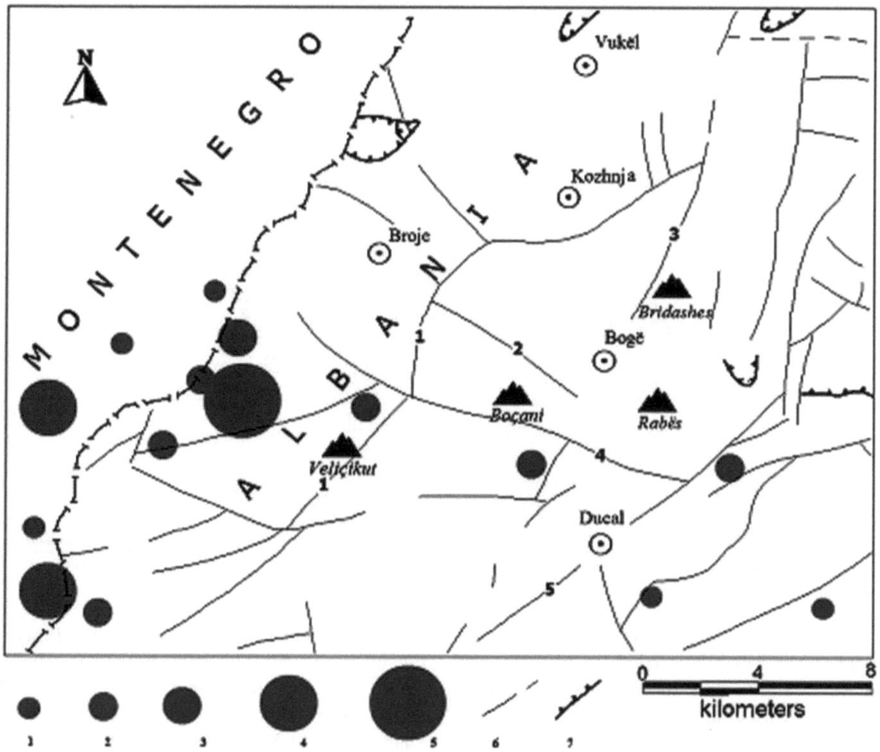

Fig. 2.24 Main fault structures and epicenters of recorded earthquakes for the period 1961–1988 in the studied area (Shanov 1999—Reproduced by permission of *Geologica Balcanica*) Earthquake magnitudes: *1*—*M* < 2.0; *2*—*M* = 2.1–3.0; *3*—*M* = 3.1–4.0; *4*—*M* = 4.1–5.0; *5*—*M* = 5.1–6.0; Faults: *6*—normal faults; *7*—thrusts; More important regional faults: Pjetroshan-Veliçikut-Kozhnje; *2*—Dobromiri; *3*—Böge- Budace; *4*—Rapsh-Boçani; *5*—Ducal

following tectonic events during the Alpine tectonogenetic phases and especially the intense overthrusting of a part of the limestones complex to the South increased the thickness of the rock complex (Fig. 2.26).

The neotectonic uplift and denudation led to the contemporary complex relief, with altitudes exceeding 2,000 m. It could be supposed that the thickness of the carbonate rocks, now subjected to karstification, is more than 1,500 m. An argument for this supposition is the concentration of important karstic springs at altitude of some tens of meters over sea level and running out of impressive quantities of fresh water in Shkodra Lake and Adriatic Sea. The tectonic factor should be also taken into account—fault sets that facilitate the drainage of superficial water outflow by channels, which do not allow always the formation of accessible karstic cavities.

According to the Hydrogeological Map of PSR of Albania M 1:200 000 (1981), the fault marked by the line Pjetroshan-Velicikut-Kozhnje plays the role of main distributor of underground water outflow in the region localized west-northwest of Böge settlement. There are concentrations of karstic springs along the northwestern fault wall. Probably, an enormous underground water flow is drained by the same fault, coming out on the surface near the village of Pjetroshan by a well-developed karst spring system. One of the springs has a discharge up to 4 m^3/s.

At the same time, the southeastern fault wall, constituted by faulted blocks, offers all conditions needed for draining of huge superficial water quantities (most of all atmospheric) and their running toward the drainage zone. The investigated region is localized in joints-karstic type of water-bearing system having an effective infiltration ratio about 0.6–0.7. In these conditions, the atmospheric waters penetrate fast in the water-bearing system and flow toward their natural drainage level. In this way, no conditions stimulating the development of horizontal

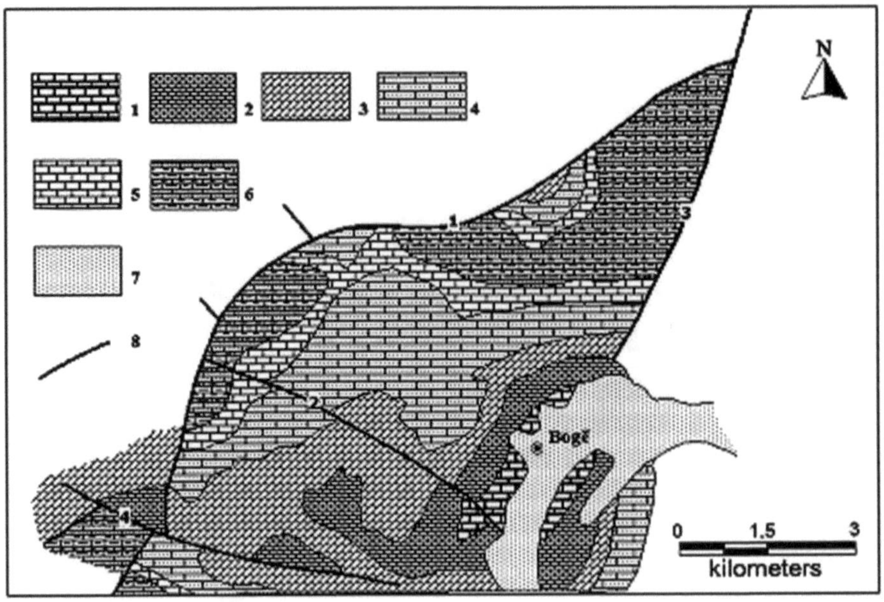

Fig. 2.25 Geological scheme of the investigated territory NW of the Böge settlement (North Albania), carried out on the base of Geological Map of Albania in scale 1:200 000 (1981) (Shanov 1999—Reproduced by permission of *Geologica Balcanica*) *1*—Upper Triassic limestones and dolomites; *2*—Lower Jurassic limestones and dolomites; *3*—Middle–Upper Jurassic undivided limestones and dolomites; *4*—Upper Jurassic limestones and clayey sands; *5*—Tithonian limestones; *6*—Lower Cretaceous limestones, carbonate limestones, clayey sands; *7*—Quaternary sediments in superimposed depressions; *8*—faults: *1*—Pjetroshan-Velicikut-Kozhnje; *2*—Dobromiri; *3*—Böge—Budace; *4*—Rapsh–Boçani

karst system could practically exist. The conditions for vertical channels precipices are much better. Several expeditions of the Bulgarian Speleological Federation just prove the mass development of precipices reaching depths from tens to 500 m (Fig. 2.27).

The speleological as well as the geological structural investigations in the karstic massif to the northwest of Böge settlement (Fig. 2.23) clarified additionally some problems and showed a strong significance of the fault tectonics as a factor for precipices and caverns formation in the most fragmented by young displacements zones.

The measurements of the space elements of joint systems and tectonic striations on different surfaces of outcrops have been carried out only at 16 points (Fig. 2.28) because of relief complexity, its difficult access and the limited time for field observations. The information turned to be sufficient for reconstruction of the main tectonic deformation phases after Early Jurassic times up to present days, which determine the main relief characteristics as well as the superficial and underground karst features.

2.3.1.3 Investigation Methods

Three basic methods were used for reconstruction of the principal axes of the tectonic stress field. Detailed analysis of the final results from each observation point allowed to distinguish the different deformation phases imprinted in the structure of the joint systems of the karstic massif and controlling the basic characteristics of the karstic processes and to determine their age.

Fault-Plane Solutions from Earthquakes

Only one fault-plane solution was available for the purposes of the investigation, i.e., an earthquake with magnitude $M = 5.1$ (f November 3, 1968) and epicenter localized to the southwest of the studied region (N 42.10, E 19.35°). The data are taken from the paper of Muço (1994).

Tectonic Stress Field Reconstruction from Striations on Slickensides

These investigations occupy the most significant place in the collected information because, in 11 points of

Fig. 2.26 Highly deformed and fractured limestones forming the borders of the valley of Böge

structural studies, indicators of displacements on the joint and fault surfaces have been found—exceptionally well-preserved tectonic striations covering in some cases surfaces of several square meters.

Reconstruction of the Tectonic Stress Fields Using Tectonic Shear Joints

This study has been carried out following the method of Nikolaev (1977), which has been already commented. In the region studied, mass measurements of joints have been carried out in four outcrops.

2.3.1.4 Results

Structural investigations applied for rock formations of different age (but within the Jurassic Period) allow to differentiate the reconstructed tectonic stresses in time. Table 2.2 shows diagrams of the reconstructed tectonic stress fields for the sites of performed measurements, which are related to a certain tectonic phase of a global scale. This turned out to be the most logical scheme when interpreting and comparing reconstructions for the sequence of Jurassic limestones of different age.

Fig. 2.27 Typical subvertical karst precipice in Albanian Dinarides

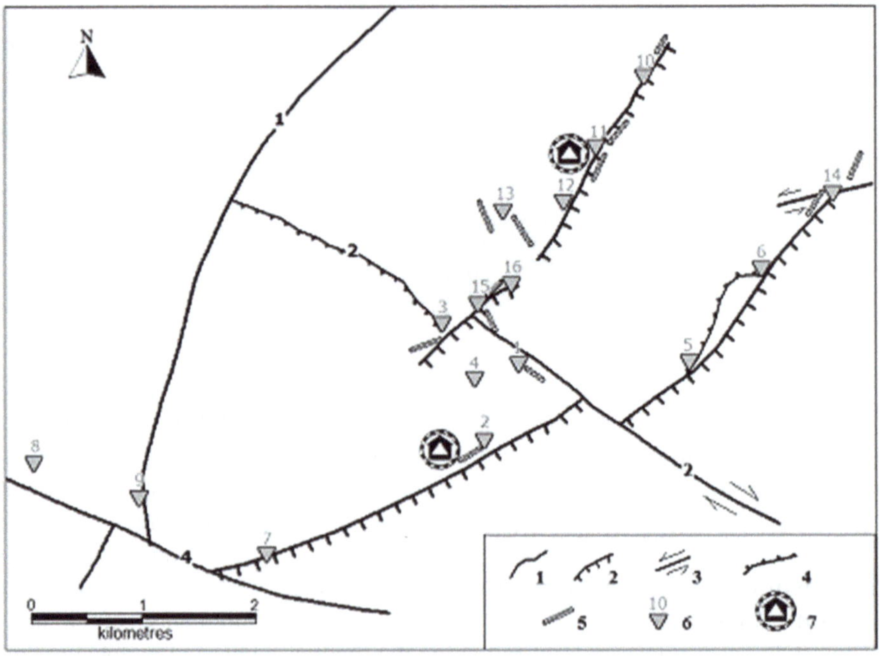

Fig. 2.28 Scheme of fault ruptures in the region investigated in 1994, sites of structural measurements and base karst forms orientations in plan view *1*—fault with undetermined type of displacements; *2*—normal fault; *3*—strike-slip fault; *4*—thrust; *5*—sites of structural measurements; *6*—speleological camps. Fault numbers follow these in Fig. 2.25: *1*—Pjetroshan-Velicikut-Kozhnjc; *2*—Dobromiri; *4*—Rapsh–Boçani

The youngest (Quaternary and recent) tectonic stress field is defined by the fault-plane solution of the only available earthquake focus near the territory studied. Its features (NW–SE compression and NE–SW extension) could be found in the reconstructions for sites 13, 14, and 9. The tectonic stress field is characterized by changed stress axes as compared to those of the Neotectonic (post-Miocene)

Table 2.2 Reconstructed time sequence of the tectonic stress fields for the studied area of Albanian Dinarides (Shanov 1999—Reproduced by permission of *Geologica Balcanica*)

Origin	Site №	TECTONIC PHASES: Late Cimmerian	Laramian	Pyrenean	Neotectonic	Recent	Method
Earthquake 03.11.1968 M=5.1 N 42.10⁰ E 19.25⁰							Fault-plane solution
J_3 - Limestones	5			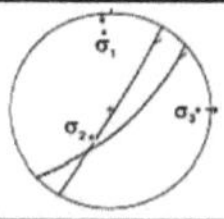			Striations on slickensides
	6				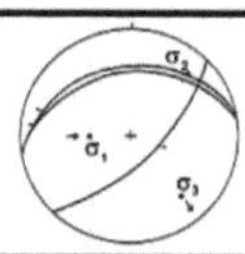		
	10				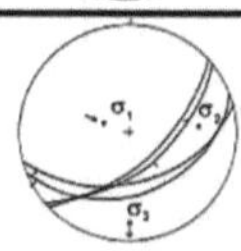		
J_3 - Limestones	12				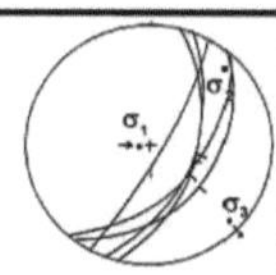		
	13		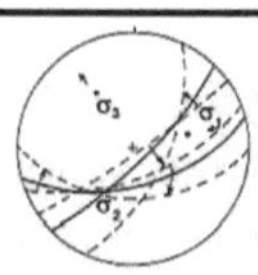	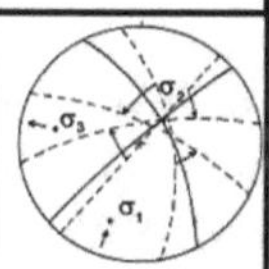	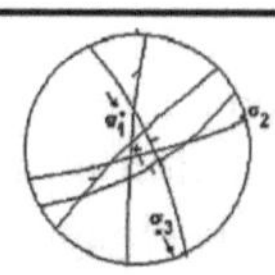	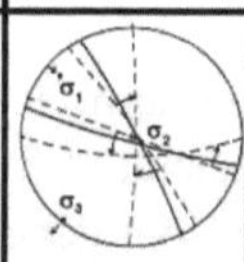	Striations & Shear joints
	14		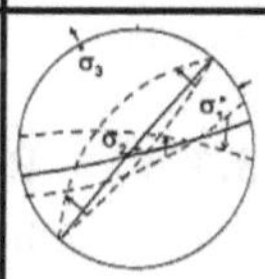			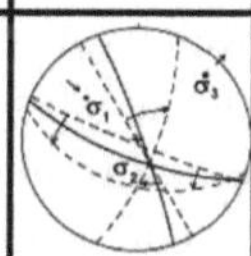	Shear joints
	15		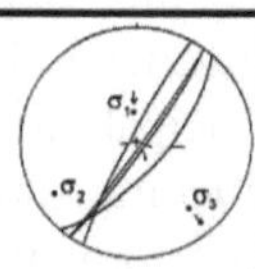				Striations on slickensides
	16		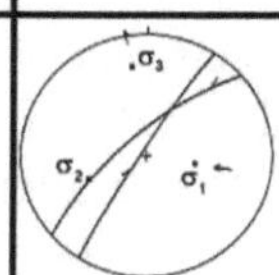				

(continued)

Table 2.2 (continued)

$J_{2\text{-}3}$ - Limestones	4	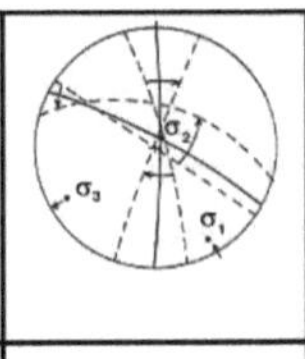		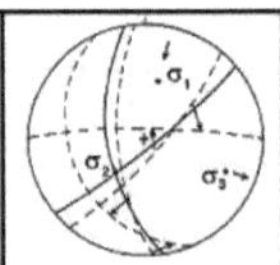			Shear joints
	7				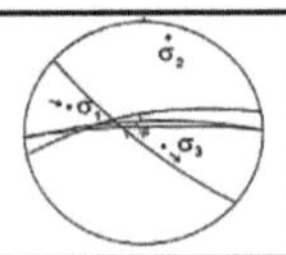		Striations on slickensides
	9					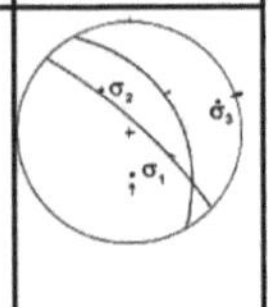	
J_2 - Limestones	2	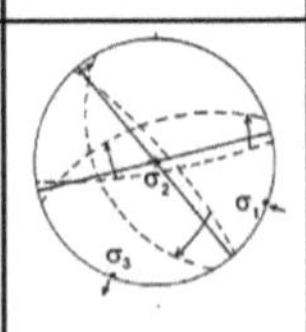	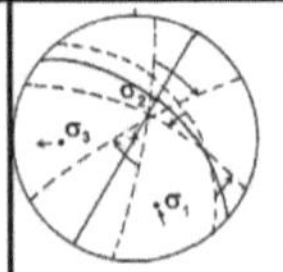				Shear joints
	8	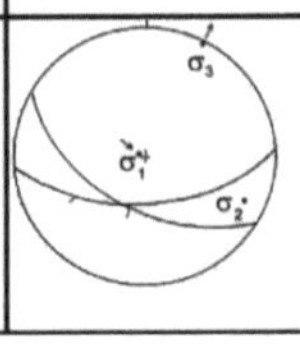					Striations on slickensides

stage, where the minimum tectonic stress σ_3 of NW–SE to E–W trend generated formation of neotectonic superimposed depressions generally elongated NE-SW (see Fig. 2.23). The compression was subvertical (uplifting) and it is well expressed by the striations on the tectonic slickenside at sites 6, 10, 12, and 13.

The most intense deformations can be referred to the Pyrenean phase (end Eocene). Then, under the conditions of strong pressure of almost N–S direction (see sites 5, 13, and 4), a southward thrusting of the Albanian Dinarides onto the Krasta-Çukali zone occurred. After the Middle Jurassic, Late Cimmerian, and after that, Laramian tectonic deformations affected the massif of limestones and dolomites. Reconstructions in sites 13, 14, 15, 16, and 2 are a clear evidence of this. The Laramian Phase is characterized by a strong subhorizontal northwest-southeast trending minimum principal stress σ_3.

The Late Cimmerian Phase shows features, which are very similar to the contemporary tectonic stress field. Taking into consideration that during the Pyrenean Phase the whole massif has been displaced to the south, presumably with some rotation, both most ancient phases posse an element of relativity in reconstructed directions of their principal tectonic stress axes.

2.3.1.5 Relations Between Karst Formation and Concrete Tectonic Conditions

The first step for determination of tectonic factors, influencing karst formation, is the analysis of the principal stresses that have led to the now observed structural ruptures in the discussed rock massif. The karst formation has started most likely at the end of the Eocene-Oligocene stage. By this time, the displacements of the limestone massif to the south had calmed down, and its intense uplifting started.

The detailed observations carried out in the sites of reconstruction of the tectonic stress field have resulted in some changes of the ideas concerning the fault network of the area. On the one hand, the normal faults are complicated by strike-slip displacements, being of the youngest age, and they correspond to the youngest (contemporary) tectonic stress field (Table 2.2). The well-expressed thrust structures (sites 3, 5, and 6) appear as relicts of the Laramian tectonic phase deformations. In the zones of shearing and mylonitization, there are conditions for formation (mainly because of mechanical export of particles by the water) of small horizontal cavities, which seldom reach more than several tens of meters (for example at sites 5 and 6). Their orientation is along the slip surface.

During the Pyrenean Phase, strong normal faulting along the ruptures of NE-SW direction and opening of joints systems into the same direction occurred. Most investigated vertical precipice caves developed on these joints. It is worth to emphasize that they all belong to zones with well- manifested faults.

Precipices near sites 10, 11, 12, and 16 are located along morphologically well-manifested fault still not shown on official maps. All these precipices are located in brecciated zones with clear tectonic slickensides and striations on them.

During the Pyrenean phase, tensile stresses varied locally from NW–SE to NE–SW direction, and the compression was oriented N–S or sub-vertically (uplifting of the massif). The subvertical compression lasted also during the Neotectonic stage when the orientation of the minimum stress σ_3 favored both the existence of earlier formed open cracks of NW–SE direction (sites 13 and 15—open karren and precipices; in case of site 1—widely open karren up to several meters). Some evidence of strike-slip displacements can be seen on site 14. The controlling role of the faults for the process of massif karstification is best observed in areas of Upper Jurassic clay sandstone, where caves could not exist without the influence of the tectonic factor. Strong faulting and crack opening favor mechanical export of debris and the formation of typical karstic precipices.

The contemporary displacements are not so active but in case of NW–SE orientation of σ_1 they have led to some clear strike-slip displacements along the Dobromiri Fault. Besides that, the erosion rate obviously becomes dominating over the uplifting process, and this results in sealing of many of the superficial karstic forms with deluvial materials. Thus, in spite of the high permeability for atmospheric waters into the karstic massif, the vertical type of open karstic systems are conserved only within fault zones, where, nevertheless, the young tectonic displacements do not allow their fast colmatation with materials from the rock weathering.

The sharp-shaped relief is a result not only of tectonic displacements, but also of chemical destruction of limestones and dolomites by rains and snow. The long lasting snow cover at this altitude (between 1,000 and 1,500 m) is a factor for chemical erosion durability through the year. Special analyses in the region for atmospheric waters aggression potential, as well as of the snow covering melting waters have not been done. Such an investigation would deepen the analysis of the reasons for the strong superficial and underground karstic processes in Albanian Dinarides.

2.3.2 Bulgaria: Tectonic Stress Fields Studies in Karst Systems

Bulgaria covers an area of 111,000 km^2. About 23 % of this area consists of carbonate rocks (i.e., pure limestone or dolomite) that host over 5,000 caves. Carbonate rocks, in which the majority of karst is developed, were formed during the Triassic, Jurassic, Cretaceous, and Tertiary periods. Less extensive karst with minor significance with respect to cave and karst development exists in Proterozoic marbles of Rila-Rhodopes area (Fig. 2.29).

Geomorphologically, there is an expressed distinction between the four principal regions (Popov 1982): (I) Danube Plain (Moesian Platform) in the northern part of Bulgaria; (II) Stara Planina (Balkan) Zone; (III) Intermediate Zone occupying the central parts of the country; (IV) Rila-Rhodopes Zone in the south (Fig. 2.29). The boundaries between the regions are roughly coinciding with faults and fault segments activated during the Neotectonic stage, some of them active till now and controlling the contemporary terrain morphology (USGS 2004). Only some of the active faults are recognized by the method of Paleoseismology (Meyer et al. 2007; Vanneste et al. Vanneste et al. 2006), the others are plotted on the base of published data (Cadet and Funiciello 2004; Georgiev et al. 2007; Kastelic et al. 2011; Radulov et al. 2006, 2011, 2012; Tzankov et al. 1998; Vapcarov et al. 1974; Vrablianski 1974), or verified during field works for different projects.

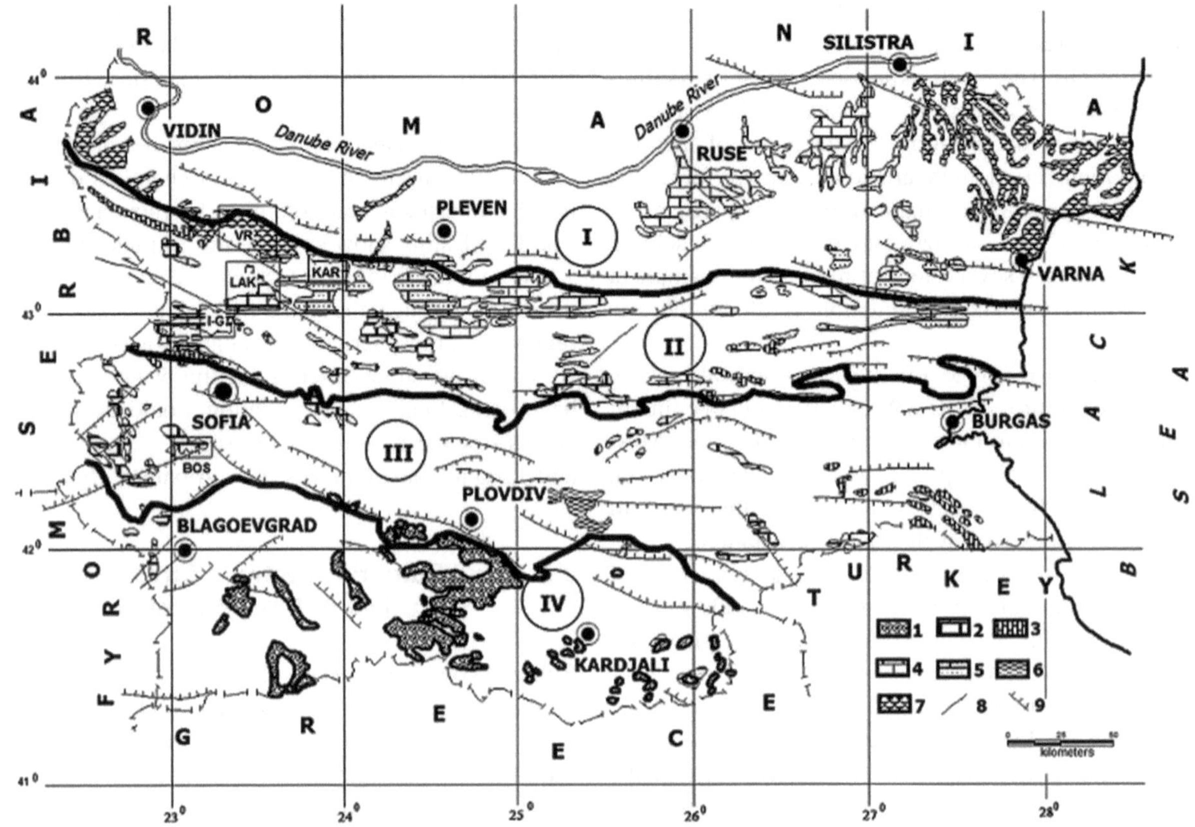

Fig. 2.29 Outcrops of karstified rocks on the territory of Bulgaria and principal karst zones, according to Popov (1982) Main regions: *I*—Danube Plain (Moesian platform); *II*—Stara Planina (Balkan) Zone; *III*—Intermediate Zone; *IV*—Rila-Rhodopes Zone Karstified rocks: *1*—Proterozoic marbles; *2*—Triassic limestones; *3*—Jurassic limestones; *4*—Upper Cretaceous limestones; *5*—Maastrichtian limestones; *6*—Paleogene limestones; *7*—Sarmatian limestones; Faults: *8*—strike-slip or undetermined type of movement; *9*—normal

The regions differ with respect to their cave and karst development, and on these grounds each of them was further subdivided into several karst districts. The Danube Plain has been discussed in Sect. 2.2 by the study of the area of Orlova Chuka Cave amd examples from the other regions are also presented .

2.3.2.1 Western Balkan Mountain

Western Balkan Mountain (Western Bulgaria), part of Stara Planina Zone, is known with important karst systems draining significant quantities of fresh water. This is an expressed mountainous area. The altitudes of the main ridge vary between 1,000 and 2,000 m, and the lowest part in Iskar River gorge is at 300–400 m. From hydrogeological view, six independent karst regions are distinguished in Western Balkan Mountain (Benderev et al. 2005). Case studies will be presented for three of them: VR—Vratsa karst region; LAK—Lakatnik karst region; I-G—Iskrets-Gubesh karst region. Their relative position is marked in Fig. 2.29.

Vratza Karst Region (VR)

Vratsa Karst Region, situated in the northern part of Western Balkan Mountain, is formed in deformed limestone layers of Jurassic and Lower Cretaceous age. The faults have been of important role for the development and evolution of the karst system. Studies on the tectonic stress fields (Fig. 2.30) were performed principally on Vratsa Block during speleological expeditions in 1985 and 1986 (analyses of the fracturing) and the preliminary results were published later (Chanov 1988). New data were included in this study, based on studies of striations on slickensides, electrical anisotropy, as well as reference on fault-plane solution from an earthquake in the

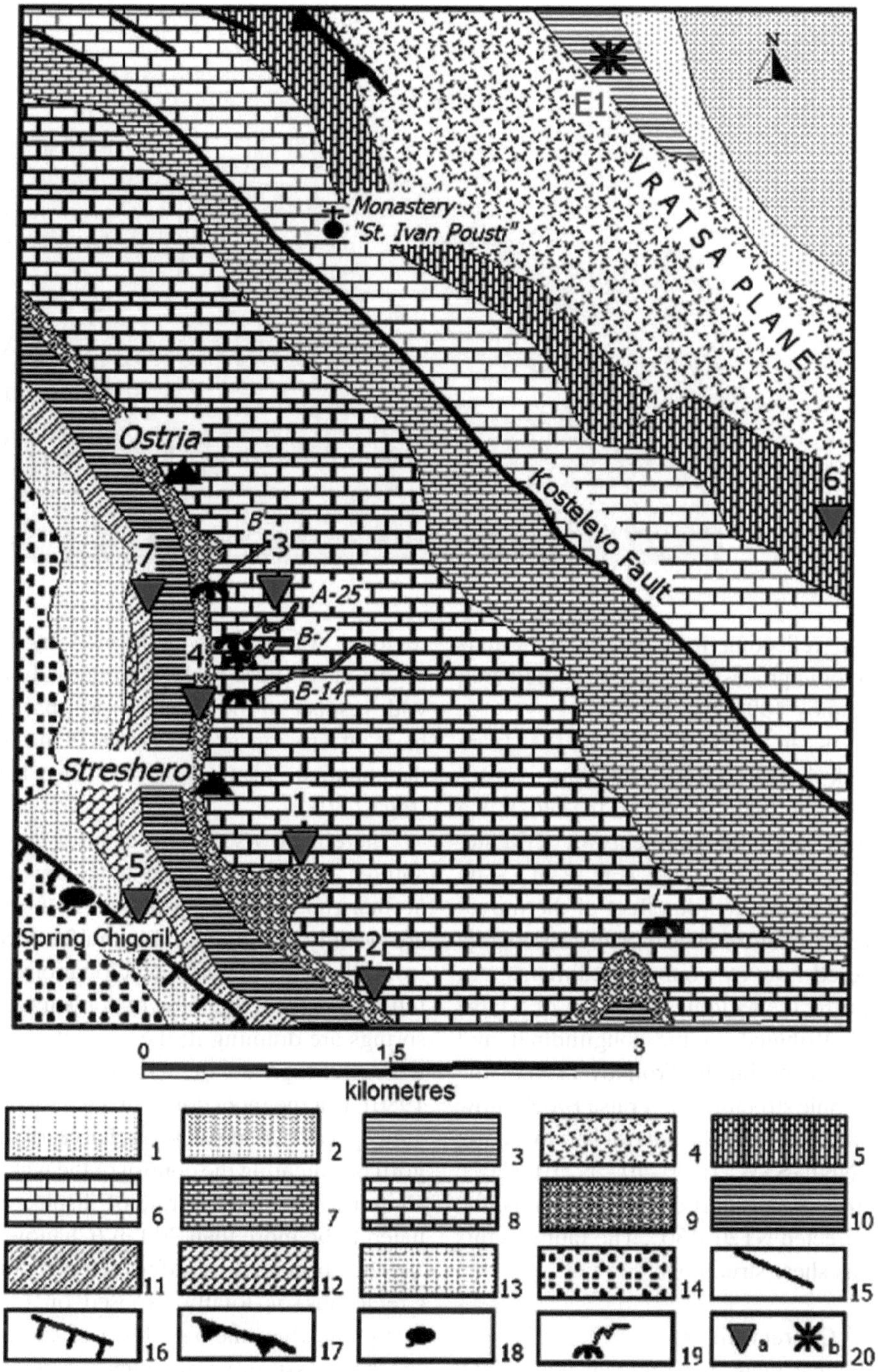

Fig. 2.30 Geological map of the area of Vratsa Block based on the Geological Map of Bulgaria, scale 1:100 000, sheet Berkovitsa (Haudutov and Dimitrova 1992) and Tronkov (1965) with the sites of reconstructions of the tectonic stress field and the principal caves. *1*—aQh: alluvial deposits; *2*—prQ_{p-h}: proluvial deposits; *3*—aQp: alluvial deposits of the older terraces; *4*—dQp: deluvial-proluvial sediments; *5*—lK_1^{ap}: limestones, marles and carbonate sandstones (Lyutibrod formation); *6*—$čK_1^{ap}$: Urgonian limestones (Cherepish formation); *7*—$slJ_2^{t} - K_1^{v}$: limestones of Slivnitsa formation; *8*—$XXJ_2^{c} - K_1^{bs}$: micritic limestones of Yavorets formation; *9*—pJ_2^{bj-b}: bioclastic limestones (Polaten formation); *10*—e/stJ_2^{a-bj}: black argillites of Stefanets Member; *11*—XIV $J_1^{h} - J_2^{a}$: sandy bioclastic limestones of Kostenets formation; *12*—XII $T_2^{1} - T_2^{k}$: dolomites, marls and argillites from Beli Izvor formation; *13*—PeT_1: Petrohan Terrigenous formation; *14*—vr P_1: Permian breccias and conglomerates of Vrana formation; *15*—fault; *16*—normal fault; *17*—overthrust; *18*—spring; *19*—caves with horizontal projection of the main galleries: *L*—Ledenika Cave; *B-14*—Barki 14; *B-7*—Barki 7; *A-25*—Academic 25; *B*—Beliar; *20*—**a** site of study of the fracturing; **b** site of study of the electrical anisotropy

Fig. 2.31 View on the eastern boundary of Vratsa Block, Bulgaria

Moesian Platform, reflecting the contemporary tendencies of the tectonic processes near the studied area (Georgiev and Shanov 1991).

Vratsa Block is an integral part of the large Berkovitsa Block-Anticlinorium structure (after Tronkov 1965). The northern border of the block is a remarkable flexure, built of sediments of Triassic, Jurassic and Early Cretaceous age (Fig. 2.31). The flexure is discussed as the ductile effect of the movements along a large fault, located northwards of the flexure, and covered now by younger sediments. An important role during the tectonic evolution of Vratsa Block has been attributed to the longitudinal and transversal faults. According to Tronkov (1965), the analyses of all tectonic structures of Vratsa Block show their genetic relationship to the lateral strain acting with direction NNE–SSW (N30–40—N210–220°). This is the direction of the short deformation axis, the long axis being directed N120–130°. The faults in this situation appear as shear structures.

Hydrogeological Characteristics

From hydrogeological view, the regional geological, tectonic, physical, and geographical conditions predestined the presence of a totally drained monoclynal slope dipping toward north with two principal karst aquifers—Triassic and Middle Jurassic. A relatively thin terrigenous rock complex (Lower and Middle Jurassic) is dividing the two aquifers. The common regional water basis is formed by Lower Triassic sandstones. These rocks are relatively highly elevated with respect to the local erosion basis. The northern basis of erosion is the Vratsa Plane, separated from the studied karst area by the Kostelevo Fault (Fig. 2.30).

There is now sufficient data for the Lower Triassic aquifer. It is represented by a few outcrops in the northern and eastern peripheries of the basin. The rocks are weekly karstified. The aquifer receives its water from the atmospheric precipitation, as well as by infiltration from the lying on top karstic aquifer. A few springs are draining it; the biggest of them is Chigoril with discharge rate of 7–11 l/s (Antonov and Danchev 1980). But the more constant and independent from the atmospheric conditions discharge of the spring Chigoril is indicating the control of the water by the fault of Chigoril. The vertical rate of displacement was evaluated to be more than 300 m (Chanov 1988).

The principal aquifer (Upper Jurassic—Lower Cretaceous) is totally exposed on the surface. This fact predestines the high level of karstification related to the possibility of intensive feeding from the atmospheric precipitation. According to Spasov et al. (1998), more than 54 % of the atmospheric precipitation (in average 1,000 mm/m^2 per year) is feeding the underground waters. Part of the water forms temporal streams on the surface, but the water is quickly drained in the dolines. The karst springs situated in the northern part of Vratsa Mountain are draining the aquifer. Most of them are of ascending type.

Characteristics of the Karst of Vratsa Karst Region

The rainfalls are one of the most important factors determining the active development of the karst processes in the region. The classic type of karst is represented in the upper part of the mountain (Scorpil and Scorpil 1895, 1898; Radev 1915; Mishev and Popov 1958: Markowicz et al. 1972; Kostov 1997). Its formation is also controlled by tectonic factors, as well as by the geomorphological evolution of the area (Ilieva et al. 1981; Angelova et al. 1995, 1999).

Vratsa Karst Region represents a plateau with steep, even vertical slopes, except the narrow band from the south connecting it with the adjacent karst basin. It is built by a north-northeast dipping monoclyne of limestone layers of Late Jurassic and Early Cretaceous age (Fig. 2.30). They are separated from the underlying Iskar Carbonate Formation by a thin layer of non-karstified rocks of Early Middle Cretaceous age. The principal erosion basis is in the northeastern part of the plateau. The principal karst springs appear at that place. But, as a general rule for the entire area, the boundary between the karstified and non-karstified rocks is above the level of the local river network.

The principal part of the basin is characterized by outcropping on the surface karst with abundant karst forms. Some of the dolines are with dimensions 1–2 km^2, the biggest one reaching 2–2.2 km^2. The underground karst forms are predominant. More than 70 caves and precipices are known in this area, most of them in the limestones of Late Jurassic-Early Cretaceous age. Some zoning can be traced from southwest to northeast. The southwestern part contains relatively little horizontal caves. Above them, near the peaks Streshero and Ostria, there are considerable and complicated as morphology precipices. Here are the deepest Bulgarian caves Barkite 14, the caves Beliar, Barkite 8, Mijishnitsa. All of them begin from dolines and continue as a system of steps or dipping generally toward northwest segments, following the upper boundary of the underlying karst resisting rocks. Little water streams exist in every one of the caves, their discharges are depending directly from the atmospheric precipitation. The horizontal projections of these caves on the surface are shown in Fig. 2.30.

Typical precipices of different depths characterize the zone along the northeastern border of the plateau. The deeper one is Haydoushka Precipice, near the Ivan Pousti Monastery. Its entrance begins with a 108 m deep well. Horizontal caves are also known, some of them with springs. These ones are grouped near the Saint Ivan Pousti Monastery, the longest cave being of 546 m, and with a water stream inside. The tectonic control of karst along the border of the plateau is well expressed near Ivan Pousti Monastery (Fig. 2.32).

The seven longest caves of the region include 75 % of the total length of the galleries. The ratio between total vertical to total horizontal lengths shows the domination of the vertical karstification over the horizontal.

The karst formation began during the Early Miocene and continued during Pontian and Pliocene time. When Popov (1964) studied the genesis of Ledenika Cave, he showed that its formation dates from the beginning of the Dacian time, and it is related to the Pontian level of denudation. The same period of formation can be attributed to the caves and the precipices in the high parts of the plateau.

Results from the Tectonic Stress Fields Reconstructions

The reconstruction of the tectonic stress fields was the first step of the study. The second one was to adjust the received solutions in a tectonically logic scheme for the time of manifestation of a given tectonic strain. This scheme (Table 2.3) reflects the possible temporal evolution of the tectonic stress axes orientation for the area of Vratsa Block.

Looking at the performed reconstructions of the tectonic stress fields, the most coinciding with the described above tectonic deformations in macro scale are these from the method using the shear joints.

The oldest deformations are detected on site No 5 where the Triassic limestones retain the "memory" for two impacts. The first one is related to the Early Cimmerian Tectonic Phase (at the end of Upper Triassic time), and the second—to the Late Cimmerian Tectonic Phase (the second half of Upper Jurassic time). The limestones of Polaten Formation (site No 4) were also deformed during the Late Cimmerian Tectonic Phase. A second tectonic stress field is reflected in the fracturing, and it was reconstructed on sites No 1, 2, 3, and 6.

Fig. 2.32 Expressive tectonic control on the karst along the southern boundary of Vratsa Block near Ivan Pousti Monastery, Bulgaria

We suppose that this tectonic stress field is related to the Sub-Hercinian Tectonic Phase (at the end of the Turonian). On these three sites, a second group of clearly expressed joint systems exists, assumed to be a result of the Pyrenean Tectonic Phase (end of the Eocene). The importance of the deformations from the Sub-Hercinian Tectonic Phase is reflected also as well expressed orientation of the ellipse of anisotropy of the Lower Cretaceous limestones (Vertical Electrical Sounding at the site E1). Less expressed is the anisotropy ellipse for the upper part of the electrical profile. The long axis of the ellipse could be accepted as an indication of younger deformations, but arbitrary attributed to the Pyrenean Tectonic Phase.

Only one of the sites of the measurements (No 6) has shown deformations that could be related, with some level of probability, to the Neotectonic stage. The argument for this is the similarity of the reconstructed tectonic stress axes directions to the result of the fault-plane solution from the earthquake with magnitude $M = 3.6$ northwards of the studied area. But, all analyzed data do not show evidence for cardinal changes of the principal directions of tectonic strain after the Pyrenean Tectonic Phase.

Table 2.3 Reconstruction of the sequence of tectonic impact on the rocks of Vratsa Karst Region

Origin	Site №	TECTONIC PHASES					Method
		Early Cimmerian	Late Cimmerian	Sub-Hercinian	Pyrenean	Neotectonic and recent	
Earthquake 23.03.1987 M=3.6 N 43.88^0 E 23.56^0						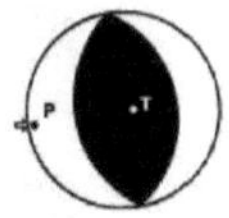	Fault-plane solution
$Q + K_1$ El	(VES)			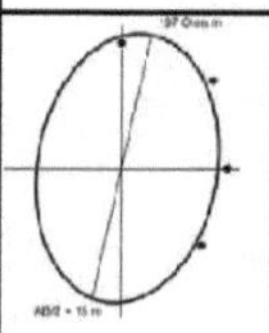	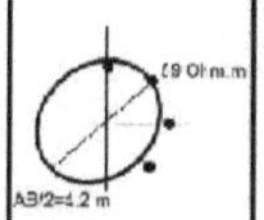		Electrical anisotropy
K_1 Limestones	6			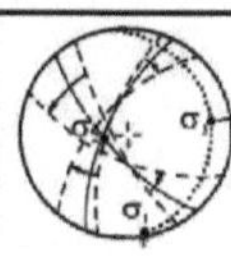	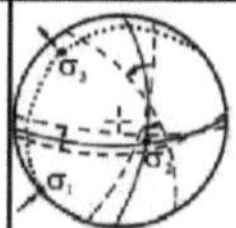	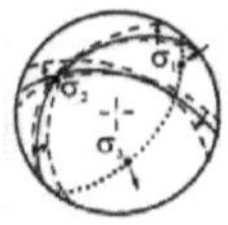	Shear joints
J_3 - Limestones	1			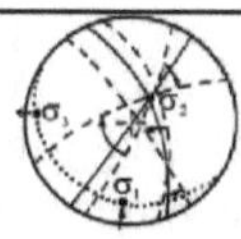	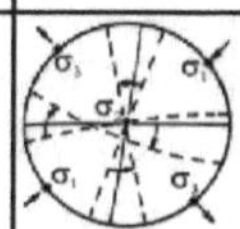		Shear joints
	2			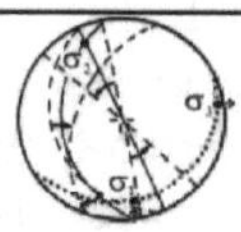	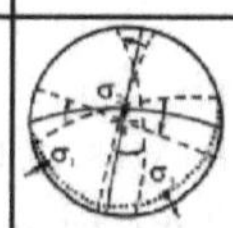		
	3			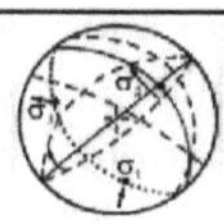	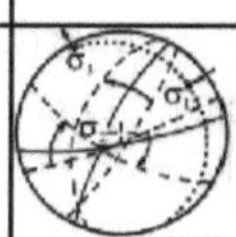		
J_2 Limestones	4		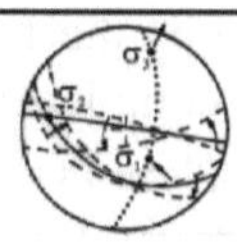	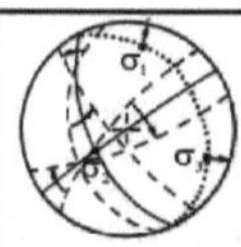			Shear joints
J_1 Sandstones	7				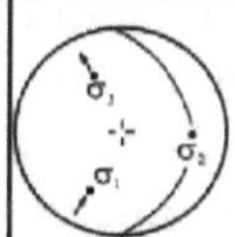		Striations on slickensides
T_2 Limestones	5	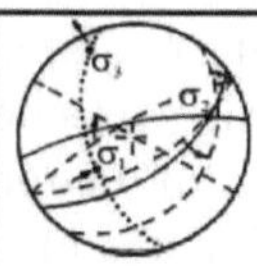	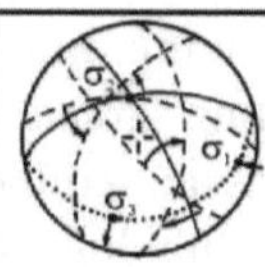				Shear joints

Fig. 2.33 Karst spring "Jitoliub", Milanovo Karst Basin, Lakatnik Karst Region, Bulgaria

This fact is reflected in the general NE–SW orientation of the karst galleries in the region. Analyzing the changes of principal stress axes from the Sub-Hercinian to the Pyrenean Phase, it is possible to deduce left rotation of the structures. For the most recent processes such a deformation was not established. The karst systems of the area are of precipice type, actually active, and draining the superficial and underground waters toward northeast. The contemporary configuration of the karst systems is possible only when the minimum tectonic stress axis σ_3 is nearly vertical.

Lakatnik Karst Region (LAK)

The Lakatnik karst spring "Jitoliub" is discharging waters drained from the southeastern part of the relatively higher karstified zone (LAK in Fig. 2.29) at the source neighborhood (Fig. 2.33)—Milanovo Basin. One of the biggest caves in Bulgaria—Temnata Dupka (more than 5 km long) is integrating the old karst levels, as well as the present active galleries and their natural water outflow—the spring "Jitoliub".

The necessity to use this water leads to ecological problems of the water qualities. One of the most important problems in the area is the correct recognition of the sites of the potential pollutant of the karst waters. A study on this problem in the context of the tectonic control of the underground cave galleries was published (Benderev et al. 2001).

Geological and Hydrogeological Background

Milanovo Basin (Fig. 2.34) with a surface of 101 km^2 is one of the biggest karst basins in Western Balkan Mountain (Benderev et al. 1987). Plakalnitsa Fault separates the basin on the northeast from the karst basins of Vratsa Karst Region. The rate of total vertical displacement along the fault is 500 m, and it disturbs the hydrological discontinuity of the karstified complexes of aquifers. The other borders are the abrupt slopes from south and southeast toward Iskar River, from southwest toward Proboynitza River, and Varshets and Ozirovo valleys from northwest. A flat plateau with average altitude of about 1,200 m is the highest part of Milanovo Basin, dominating the Iskar River valley (altitude 300–350 m).

Milanovo Karst Basin is build of Mesozoic rocks, overlying different in age Paleozoic basement (dominating granodiorite—granite complex) and Lower Triassic sediments of Petrohan Terrigenous Group. The lithological features of the Upper Triassic rocks determine the existence of one thick complex able to be karstified (Fig. 2.35), separated partially by the carbonate-terrigenous materials of Babino Formation

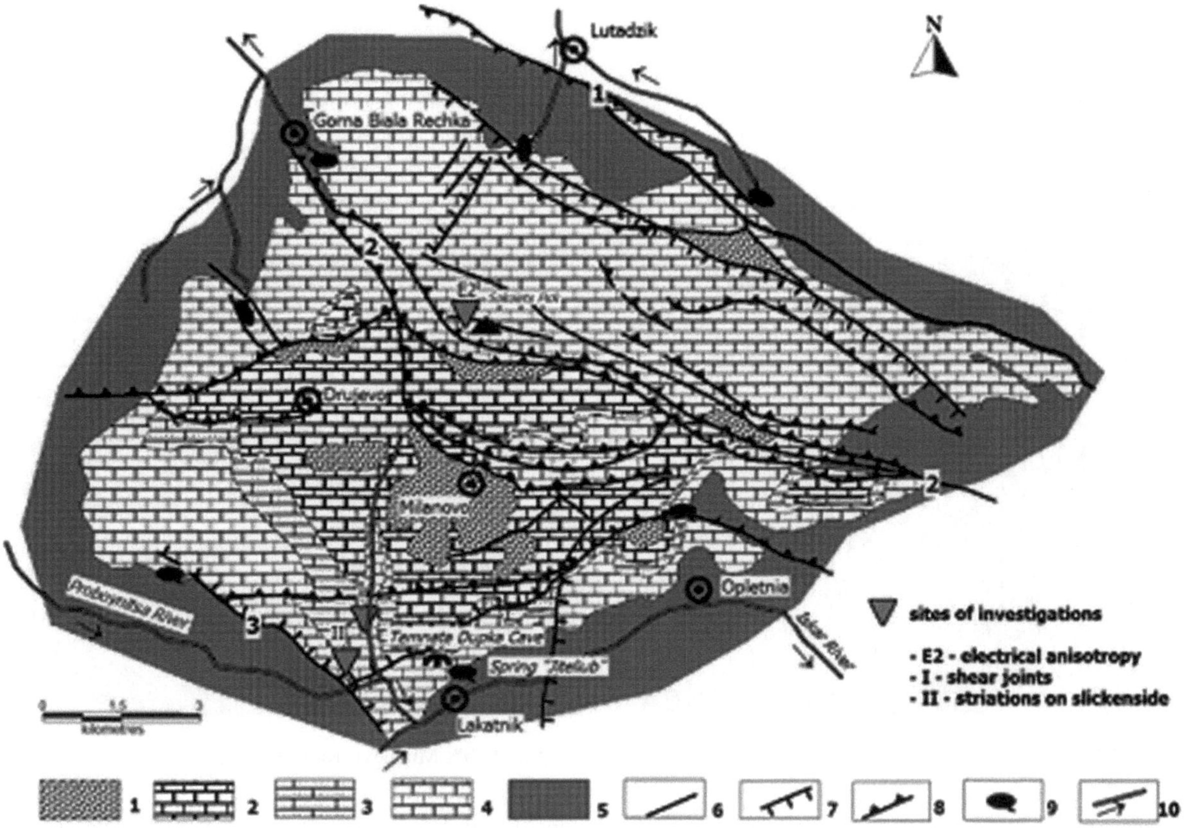

Fig. 2.34 General geological scheme of Milanovo Karst Basin (after Benderev et al. 2001) *1*—Terrigenous and carbonate-terrigenous rocks of Komshtitsa formation and Kostina formation; *2*—*Upper part* of Triassic carbonate complex (Rusinovdel formation); *3*—Intermediate Triassic carbonate-terrigenous layer (Babino formation); *4*—*Lower part* of Triassic carbonate complex (Svidol and Mogila formations); *5*—Paleozoic rock basement (granite complex) and Petrohan Terrigenous Group; *6*—fault of unknown type of movement; *7*—normal fault; *8*—thrust; *9*—Karst springs; *10*—Rivers; Main faults: (*1*) Plakalnitsa, (*2*) Pop-Sokolets, and (*3*) Proboynitsa

in two sub-complexes of limestones and dolomites (Mogila Formation with thickness of 150–350 m composed of Opletnia and Lakatnik Members, and Milanovo and Rusinovdel Formation—mainly dolomites). The lowest part of the sequence is most susceptible toward karstification processes.

The upper part of the Triassic rocks profile is composed of various terrigenous and carbonate-terrigenous rocks of Komshtitsa Formation and partially by the locally presented Jurassic terrigenous rocks of Kostina Formation.

Milanovo Karst Basin contains 103 caves and precipices, the horizontal or inclined caves being predominant. The ratio of the total altitude change to the length of the karst galleries is 0.07, and it presents the domination of the horizontal karstification. The velocity of the underground karst hydrological system hollowing is lower than the erosion on the surface and the entrenchment of Iskar River. For this reason, the lowest point of water discharging of Milanovo Karst Basin, the spring "Jitoliub" near Lakatnik Railway Station, is 10 m above the level of Iskar River. Milanovo Basin is receiving its water from the atmospheric precipitation, as well from the local rivers. Water losses were detected at the riverbed of Proboynitza River and some of its tributaries. The average total discharge of the basin springs is 1,341 l/s, or presented as a module of the groundwater flow—13.3 l/s.km^2. The minimum discharge is 197.5 l/s. The important part of the natural resources belongs to the spring of "Jitoliub" draining Lakatnik Karst Region. Some quantity of this water is used for supply of potable water.

The area near the railway station of Lakatnik is known with the largest quantity of underground karst

Fig. 2.35 View of the lithostratigraphic sequence above the spring "Jitoliub", Milanovo Karst Basin (Bulgaria)

forms—62 % of the caves of Milanovo Karst Basin. Most of the caves are not very long, but the largest cave is the only active one in the area—Temnata Dupka Cave with more than 5 km of galleries. This cave is situated at about 40 m just above the spring of "Jitoliub." The cave is a complicated labyrinth type system with underground river running out its waters trough the spring "Jitoliub." The spring water is a mixture between the underground flow from Proboynitza River and the atmospheric precipitation drained by Temnata Dupka Cave from the whole catchment zone. Analyses show, that when the spring discharge is low, the rate of feeding from Proboynitza River is less than 50 %, and it decreases with increasing spring discharge.

Tectonic Stress Fields and Preferential Directions of the Karst Galleries

Three methodologies were used for reconstruction of the tectonic stress axes (Benderev et al. 2001). The first one is based on the analysis of conjugated shear joints systems (Nikolaev 1977) and the direction of dispersion of their maximums. The measurements performed at site I in Fig. 2.34 (*I a and I b* diagrams on Table 2.4) have been analyzed. The second methodology is based on the kinematics characteristics of the movements on the rock block bounding surfaces—striations on slickensides. Clear traces of movements were found at the two sites (I and II, Fig. 2.34), and the analyses showed practically the same solution for the tectonic stress field characteristics. The summarized solution is plotted in Table 2.4 (diagram *II*).

Electrical anisotropy studies were performed near Sokolets Peak (site *E2*—Fig. 2.34). The ellipses of anisotropy were constructed using two arrays—$AB/2 = 12.5$ m and $AB/2 = 80.0$ m. The short array gives information for the first 10 m, the long array—for a depth of about 50 m below the surface. There is a clear coincidence of the long axes of the electrical anisotropy with the horizontal projections of the σ_1 axes of the two reconstructed tectonic stress fields (Table 2.4, diagrams *E*2 *a* and *E*2 *b*). The limestones are from the Triassic Carbonate Complex (T_2) and as a whole they are mostly affected by the Late Cimmerian Tectonic Phase. The main tectonic faults of the area strike NW–SE, likewise the axis of the anisotropy for array $AB/2 = 80.0$ m.

Table 2.4 Reconstruction of the sequence of tectonic impact on the rocks of Lakatnik Karst Region

Origin	Site №	Late Cimmerian	Pyrenean	Neotectonic and recent	Method
T_2 limestones	E2 (azimuthal measurements)	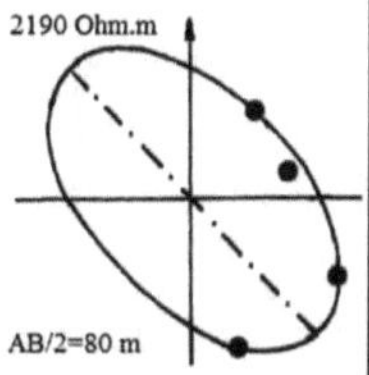 a	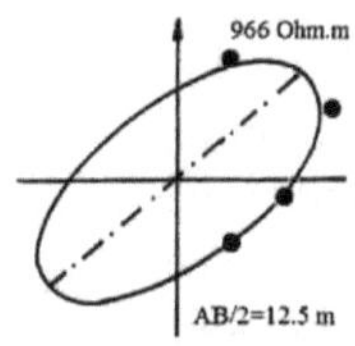 b		Electrical anisotropy
T_2 limestones	I	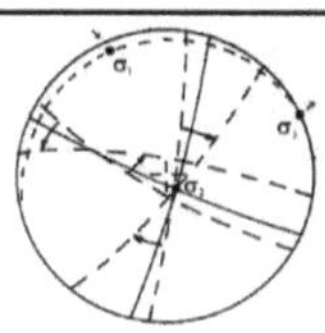 a	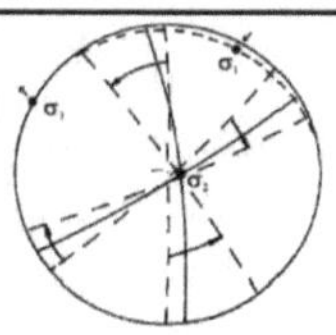 b		Shear joints
T_2 Limestones	II		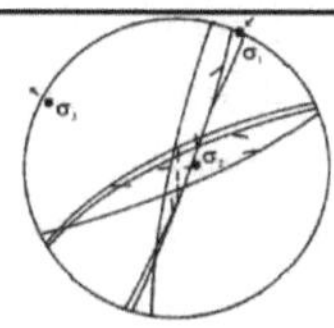		Striations on slickensides
T_2 limestones Temnata Dupka Cave	N 43.0895° E 23.3831°			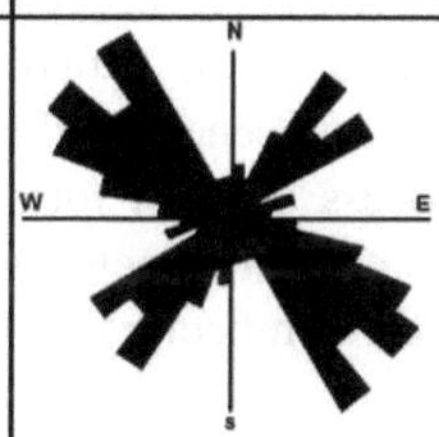	Cumulative length of the cave galleries

Better expressed is the eccentricity of the ellipse of electrical anisotropy near the surface. At site No. *E*2, the most intensive karstification is determined for the first 10–15 m. It is well seen, that the long axis of the anisotropy for $AB/2 = 12.5$ m is oriented NE–SW and it can be interpreted as reflection of the compression during the Pyrenean Tectonic Phase, realized through a number of thrusts (see Fig. 2.34), and creating conditions for opening of fractures along the same direction, and consequently their erosion and karst formation.

The conclusions about the reconstructed stress fields and the origin of the karst system are:

1. The first tectonic deformation of the whole Triassic carbonate rock complex is probably related to forces causing the overthrusting movements along the longitudinal faults of Milanovo Karst Basin (diagram *Ia* on Table 2.4). Following the studies of Tronkov (1965) it can be accepted that the deformations were a result of the Late Cimmerian Tectonic Phase at the end of the Jurassic period. It coincides with the tectonic stress field of the same age, reconstructed northward in Vratsa Karst Region (see Table 2.3, site No. 5). Conditions for opening of NW oriented fractures were created. But this is not related to the initial time of draining

of the ground waters and the formation of the NW-SE branch of the karst galleries of Temnata Dupka Cave (Table 2.4). This direction is even contemporary active and this fact suggests a coincidence of today stress axes orientations with these at the end of the Jurassic.

The inferred age of initiation of the karst formation process is 4.5 MA, i.e., from the beginning of the Pliocene (Angelova et al. 1999), and it was probably facilitated by the fractures created or reactivated during the next stage of deformations.

2. The second stage of deformation created a new system of shear joints (diagram I *b* on Table 2.4), but some of the older joint surfaces were also translators of the block movements. The striations on the slickensides confirm the existence of this tectonic stress field (diagram II on Table 2.4). This stress field favored the opening of the northeastwards striking shear joints of the older deformation stage (see diagram I *a* on Table 2.4) and their transformation in fissures. The regional studies (Chanov 1988) have given reasons for the impact of the Pyrenean Tectonic Phase on this process, as it was discussed above for the specific features of Vratsa Karst Region. It can be accepted that the same regional tectonic stress conditions had remained from the Oligocene to the Pliocene Epoch, when the process of karst formation started. The northeast–southwest directed galleries of the Temnata Dupka Cave reflect exactly this situation (Table 2.4).

Iskrets-Gubesh Karst Region (I-G)

Iskrets-Gubesh Karst Region (I-G) is the third area of Western Balkan Mountain where the geodynamic processes have a direct impact on the evolution of the karst system (Fig. 2.29). The studies, related to the karst tectonics of the area (Paskalev et al. 1992) and to the impact of the blasts from the local quarries on the discharge of Iskrets karst springs (Shanov et al. 2001; Shanov and Benderev 2005) supplied the necessary data for understanding the close relationship between the local tectonics, the tectonic stress fields and the karst features of the region.

During the twentieth century, the springs dried up several times. After Vranchea Earthquake from 1977 ($M = 7.2$) at a distance of about 400 km from the site, the discharge rate dropped down from 5.5 to 0.5 m^3/s during 7.5 h (Paskalev et al. 1992). Later, the discharge rate raised abruptly to 13.5 m^3/s, and began to decrease gradually (Fig. 2.36). Similar event happened during the local Svoge Earthquake of 1979, as well as on the April 11, 1982, when any significant earthquake was not recorded at the localities, nor in the whole Balkan Region.

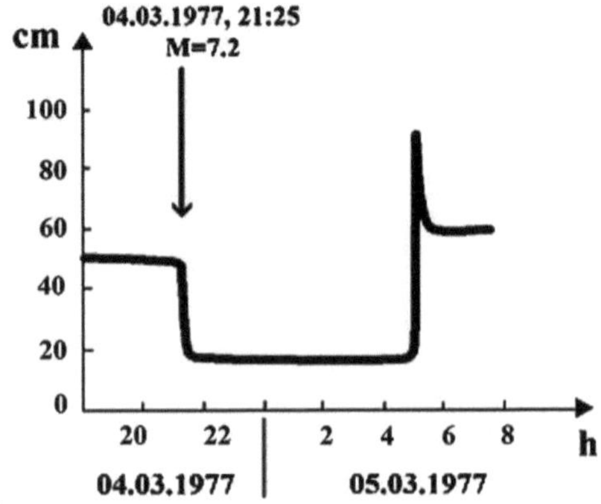

Fig. 2.36 Changes of the discharge rate of Iskrets Karst Springs after Vrancea Earthquake of 04.03.1977, M = 7.2 (Paskalev et al. 1992—Reproduced by permission of *Bulletin of Bulgarian Geological Society*)

The question for the seismic impact is of importance for the local authorities because of the significance of the springs as a major source of fresh water for the municipality of the town of Svoge. The normally used quantities are of the rate of 150 l/s. The impact from the blasts in the quarries has not the potential to disturb the normal discharge of the springs (Shanov et al. 2001; Shanov and Benderev 2005). The only factor for such disturbances can be the local tectonics and the dynamics of the processes inside the karst system.

Iskrets Karst Springs have a very variable discharge—from 280 to more than 50,000 l/s (Benderev 1989). They drained more than 80 % of the territory of Ponor Mountain (a part of Western Balkan Mountain). The springs are related to the karst system of the upper part of the Triassic Aquifer of Iskar Carbonate Group. The atmospheric precipitations on the feeding area of the springs (about 140 km^2) provide the principal quantities of water—62 % of the average annual discharge of the springs. Geological, tectonic and geomorphological conditions predestine the existence of water saturated zone north of the springs. The underground lake surface inside the neighboring Dushnika Cave has a level at 2 m above the springs' level (Fig. 2.37). The level of the

Fig. 2.37 View of the entrance of Dushnika Cave

underground lake can rise abruptly even to 13 m after intensive atmospheric precipitations and snow melting. In this case, a river appears from the lower entrance of the cave.

Geological Background

Triassic and Jurassic autochthonous sediments crop out in the area of Iskrets Karst Springs (Fig. 2.38). These rocks represent a part of the northern limb of Izdremets-Gubesh syncline. To the north of the village of Breze, it is cut by a number of E–W oriented faults. An important tectonic role is attributed to the N–S trending Ejdan-Breze normal fault (Fig. 2.38). Practically, the structures in the autochthonous sediments are controlling the underground water streams.

Svoge Allochtone is a structure built of the two regional Breze and Iskrets overthrusts. In the area of Iskrets Karst Spring, the Breze overthrust covers the sediments of the autochthone. The overthrust represents a complicated structure, including a number of overthrusted imbrications (Fig. 2.38). The plane of thrusting is almost sub-horizontal, and only its frontal part is inclined southward. The overthrust is composed of Triassic and Jurassic sediments.

Iskrets Overthrust is the most important allochthonous structure in the area. It covers and is younger than Breze Overthrust. It is built of Ordovician rocks. Folds and faults were formed in its frontal part.

The most important faults are (Fig. 2.38):

Topilia Faults. They are two subparallel faults situated northwards of the village of Breze, strike N 100–110°. The southern one is well seen in the morphology, the northern one is displacing Quaternary blockages, and this could be a reason to suppose their recent activity.

Sulio Fault is a normal fault striking N 130–140°. It continues to the east, outside the studied area. The southern block is subsided.

Ejdan-Breze Fault is a normal fault with right-lateral component of the movements. The horizontal displacement is evaluated at 250 m, while the vertical one is 40–55 m. The western block is subsided. The interpretation is (according to Paskalev et al. 1992) that the right-lateral movements appeared first and lately, during the Illyrian Tectonic Phase, the vertical displacement occurred. The fault is not displacing Topilia Faults and other northwards located faults, but it disturbs the structures of the overthrusted imbrications.

Tchernovodene Fault is one of the important tectonic structures near Iskrets Karst Springs. Its northern block is built of Babino Formation (Upper Anisian), Milanovo Formation (Ladinian) and Rusinovdel Formation (Ladinian—Middle Carnian), and the southern block is composed of the sediments of Opletnia Member (Lower-Upper Anisian). The performed

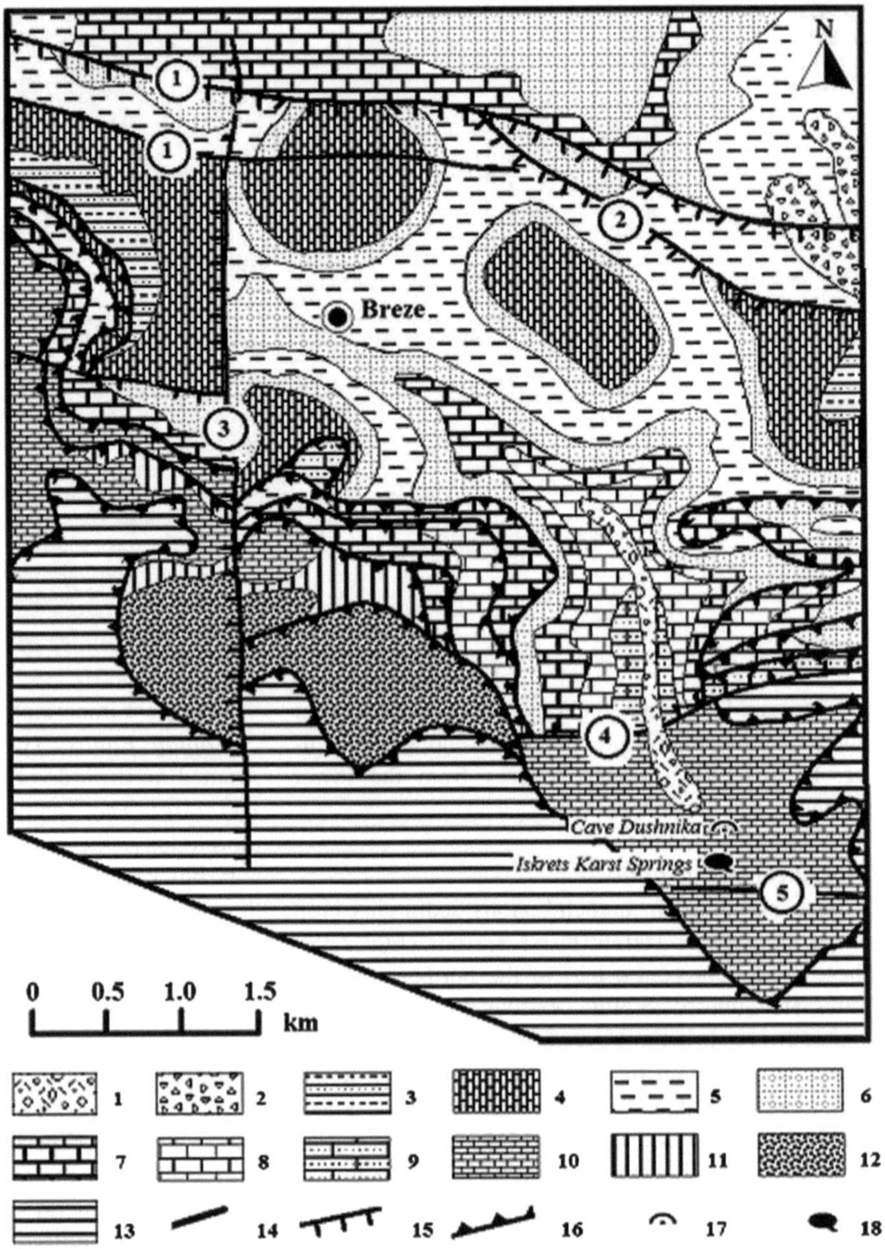

Fig. 2.38 Geological map of the area of Iskrets Karst Springs (according to Paskalev et al. 1992—Reproduced by permission of *Bulletin of Bulgarian Geological Society*) *1*—Quaternary alluvium (gravel and sands); *2*—Quaternary colluvium (rubbles); *3*—Lower Cretaceous (Salash formation); *4*—Tithonian-Berriasian (Glojene formation); *5*—Lower-Middle Bajocian (Etropole Formation); *6*—Sinemurian (Ozirovo formation); *7*—Hettangian (Kostino formation); *8*—Ladinian (Milanovo formation); *9*—*Upper* Anisian (Babino Formation); *10*—Anisian (Lakatnik Member); *11* —*Middle* Anisian (Opletnia Member); *12*—*Lower* Triassic (Petrohan Terrigenous Group); *13*—Ordovician (Grohoten formation); *14*—fault with unknown movements of the blocks; *15*—normal fault; *16*—overthrust; *17*—Cave Dushnika; *18*—Iskrets Karst springs; Faults: (*1*) Topilia faults; (*2*) Sulio fault; (*3*) Ejdan-Breze fault; (*4*) Tchernovodene fault; (*5*) Dushnika fault

geophysical studies have shown that the northern block subsided (Paskalev et al. 1992).

Dushnika Fault. It cuts the sediments of Opletnia Member, to the south of Iskrets Karst Springs. The advance of Iskrets Overthrust toward north created conditions for activation of the fault as inverse structure, and probably it created conditions for barraging the underground water flow and the formation of the springs.

Tectonic Stress Fields Reconstructions and Directions of the Karst Galleries

Two methods were applied for the analysis of the tectonic stress fields—Nikolaev's Method (Nikolaev 1977) based on the dispersion of the shear joints density and studies of the electrical anisotropy.

The shear joints analyses are based on data from three sites in the vicinity of the karst springs and Dushnika Cave (Fig. 2.39). The reconstructions gave

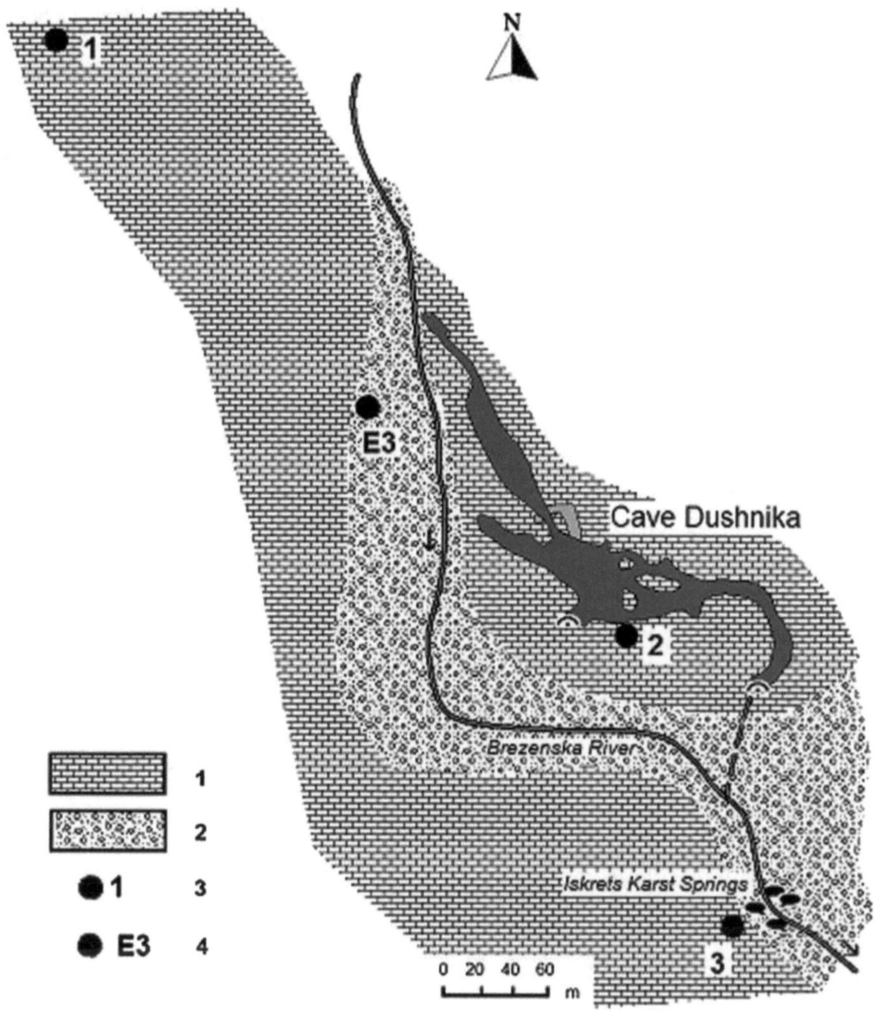

Fig. 2.39 Situation at the vicinity of Dushnika Cave with a scheme of the cave and the sites of performed studies *1*—Limestones of Opletnia Member (*Lower–Upper* Anisian); *2*—Quaternary sediments (mainly alluvial); *3*—Site of measurements of shear joints; *4*—Site of performed electrical anisotropy studies

similar solutions for all sites. The summarized diagrams using 110 measurements from the sites represent the statistically most stable solution. Two distinct tectonic impacts have been inferred that created two preferential directions of open fractures in the limestones of Opletnia Member. These two directions have the same orientations as shown for Vratsa and Lakatnik karst regions—NW–SE and NE–SW (Table 2.5). The first one is controlling these galleries of Dushnika Cave that are now active, the second is more or less coinciding with the orientation of the entrance part of the cave.

If so, the first tectonic deformation is the same that impacted the Triassic carbonate rocks in Western Balkan Mountain, and it is accepted to be a result of the Late Cimmerian Tectonic Phase at the end of the Jurassic period. Conditions for opening of NW fractures were created. These fractures obviously control the NW-SE branch of the karst galleries (Table 2.5).

The second reconstructed stress field favored the opening of the northeastwards striking fissures. The regional studies and the presented above interpretations for Vratsa and Lakatnik karst regions have demonstrated that the resulting deformations, most probably, reflect the impact of the Pyrenean Tectonic Phase, and that the same regional tectonic stress conditions had remained from the Oligocene to the Pliocene. The NE–SW oriented galleries of Dushnika Cave reflect one more time this situation (Table 2.5). The ellipse of the electrical anisotropy for the first 15 m from the surface ($AB/2 = 30$) has its long axis oriented exactly in this direction. The fault-plane solution for the nearest earthquake of 09.03.1980,

Table 2.5 Reconstruction of the sequence of tectonic impact on the rocks of Iskrets-Gubesh Karst Region

Origin	Site №	Late Cimmerian	Sub-Hercinian	Pyrenean	Neotectonic and recent	Method
Earthquake 09.03.1980 M=4.4 N 42.95^0 E 23.35^0					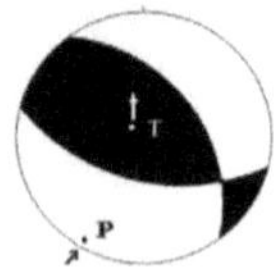	Fault-plane solution
T_2 limestones	E3 (azimuthal measurements)		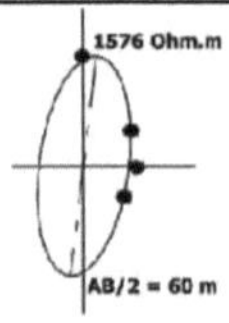 a	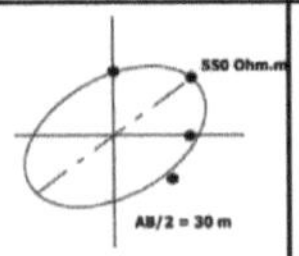 b		Electrical anisotropy
T_2 limestones	1, 2 and 3	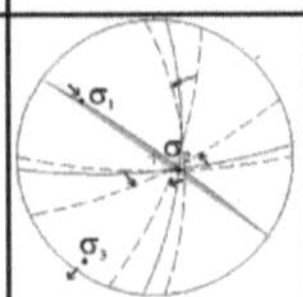 a		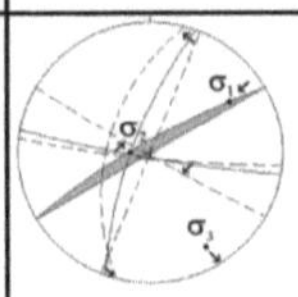 b		Shear joints
T_2 limestones Dushnika Cave	N 43.0895^0 E 23.2356^0				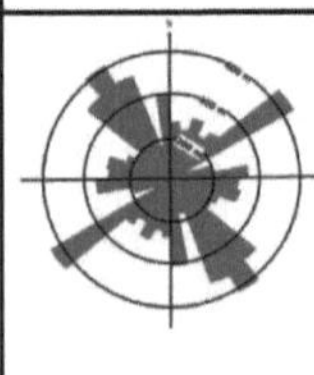	Cumulative length of the cave galleries

$M = 4.4$ (Table 2.5) gives also reason to postulate that the general orientation of the tectonic stress field has been not changed drastically since the deformational processes during the Pyrenean Tectonic Phase till now.

The orientation of the long axis of the electrical anisotropy at N 5° for depths of about 30 m ($AB/2 = 60$ m) has not any response on the diagram of the karst galleries (Table 2.5). But for Vratsa Karst Region the same orientation was detected in the Lower Cretaceous limestones and it was attributed to the Sub-Hercinian Tectonic Phase (end of Turonian time). In the studied area in the vicinity of Dushnika Cave we can only suppose that this orientation is due to the satellite fractures related to the well expressed N–S striking *Ejdan-Breze Fault*, reflected also in the morphology of the terrain as factor for the local N–S orientation of the valley of Breze River.

The cause for interruption of the underground water flow during strong seismic events is probably related to tectonic deformation or rock collapses with temporary formation of barrages inside the unknown parts of the Cave Dushnika northward of the Iskrets Karst Springs, especially in the area of Chernovodene Fault (Fig. 2.38). The existence of unknown large underground spaces has been proved by geophysical studies to the north of the known parts of the cave (Paskalev et al. 1992).

2.3.2.2 Central Balkan Area: Karlukovo Karst Region

One of the most spectacular karst areas on the territory of Bulgaria is Karlukovo Karst Region (KAR). It is located in the Central Balkan, in the morphological structure named Central Fore Balkan (Fig. 2.29). It is known especially for the cave with the spectacular

Fig. 2.40 Remote view to the entrance of Prohodna Cave, Karlukovo Karst Region, Central Balkan, Bulgaria

highest cave entrance in Bulgaria (42.5 m) of Prohodna Cave (Fig. 2.40), as well as for the multiple vertical and horizontal caves in the area.

Geological Background

Karlukovo Karst Region is situated in Lukovit Syncline. This is a well-expressed asymmetric structure in the northern part of the Fore Balkan, affecting carbonate sediments of Campanian, Maastrichtian and Paleocene age, and complicated by fold and faults (Karagjuleva 1971). The area of direct interest is situated northwards of the village of Karlukovo (Fig. 2.41). Dominant sediments are the limestones of Kailaka Formation—Late Maastrichtian; they build the westernmost part of Lukovit Syncline. Aptian and Albian sediments of Roman Formation (sands and marls) occur southward.

Maastrichtian limestones are highly fractured and karstified. This fact and the canyon type erosional incisions suggest relatively young tectonic activity, closely related to the karst processes. The faults, crossing the area of interest, are not well recognized geologically. The geophysical works during 1985–1986 (electrical profiling and WLF profiling—personal unpublished data of S. Shanov) discovered the existence of a covered fault, controlling the karst in the area.

The most significant results from the performed geophysical works were:

- It was proven that a significant fault exists, named Karlukovo Fault, striking NW–SE. The fault is natural westwards prolongation of the regional Brestnik-Preslav Flexure (Fig. 2.41);
- The fault is dividing the less karstified limestones to the north from intensively karstified limestones of the southern block. All important karst caves are located directly south of the fault;
- Karlukovo Fault offers an explanation of the hydrogeological features of the area, including the karst springs;
- New unknown karst cavities were predicted to the south of Karlukovo Fault and after a long period of exploration of Bankovitsa Cave, the speleologists discovered in 2008 new galleries with an underground river. These galleries coincide with the geophysical anomalies and their trend corresponds to the reconstructed preferred directions of the open fissures when analyzing the tectonic stress fields.

The scheme in Fig. 2.42 shows the main sites of the performed structural and geophysical studies, as well as the maps of the caves.

Tectonic Stress Fields Reconstructions

As in the above-discussed case studies, two methods were applied to analyze the tectonic stress fields—Nikolaev's Method (1977), based on the dispersion of shear joints density, and studies of the electrical anisotropy.

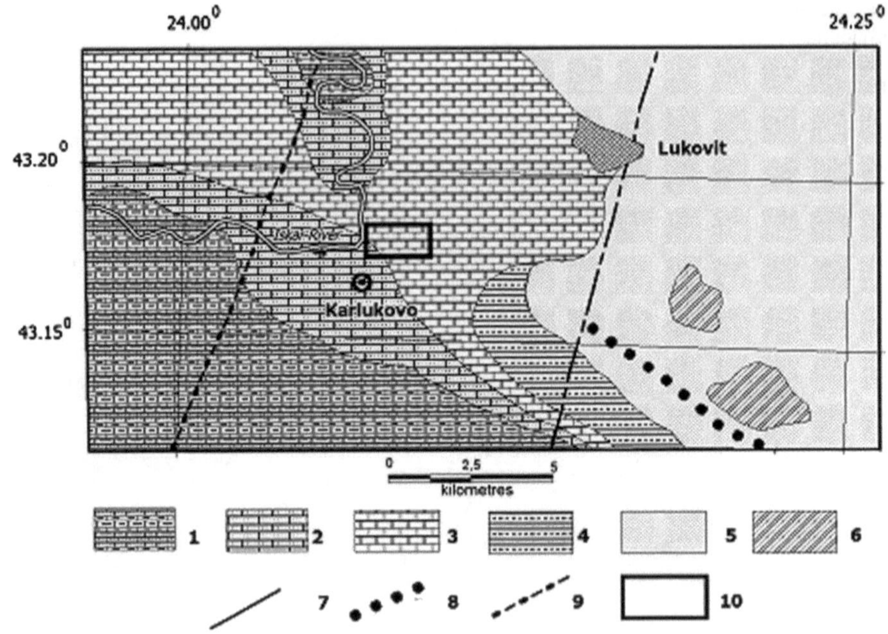

Fig. 2.41 Geological map of the region of Karlukovo Village. *1*—Roman formation rmK_1^{a-al} (sandstones and marls); *2*—Terrigenous-carbonate formation $14K_2^{cn-m}$; *3*—Kailaka formation kK_2^m (limestones); *4*—sandstones, siltstones, clays and marls $13Pg_2$; *5*—Ugarchin formation $ugPg_2$ (sandstones, clays, marls and siltstones); *6*—Opanets formation opN_1^b (clays and limestones); *7*—fault covered by younger sediments; *8*—flexure; *9*—fault line according to gravimetric data; *10*—studied territory

The summarized data on the shear joints from sites 1, 2, and 3 (302 measurements) were statistically processed and as a result two tectonic stress fields were reconstructed that had affected the Maastrichtian limestones (Fig. 2.42, diagrams *A* and *B*).

The first stress field is characterized by sub-horizontal maximum compression in NW–SE direction, while the minimum tectonic stress axis is also sub-horizontal and NE–SW oriented. This situation is favorable for opening of fissures dipping 75° northeastwards (strike N 115°). The strike is more or less coinciding with the general direction of Karlukovo Fault.

In the second stress field, σ_1 and σ_3 are also sub-horizontal, but the compression is NE–SW directed, whereas the minimum stress axis strikes NW–SE. In this case the fractures striking N 18° and dipping 80° are optimally oriented for opening and karstification.

These two fields have impacted the limestones during the Late Alpine Tectonic Phase when the main structures of the Balkan Montain were formed (45 Ma ago). The local tectonic stress conditions provoked the formation of Karlukovo Fault and the relative uplift of its southern block. The initial conditions for karst formation were created and they were more active in the southern block being exposed to more intensive denudation.

Later, probably during Neotectonic time, the general N–S compression created conditions for development of NNE–SSW fissures. The process was relatively fast and lead to the formation of canyon type valleys in the southern block, at some sites cutting the older karst forms. During Neotectonic time, the total uplift of the area is evaluated at about 300 m (Vrablianski 1974).

Studies of the electrical anisotropy were performed at three sites (Fig. 2.42) with array $AB/2 = 30$ m. Site KE 1 and KE 3 have similar solutions and the long axes of the electrical anisotropy ellipses are more or less corresponding to the strike of Karlukovo Fault, i.e., they reflect the first tectonic stress field. Site KE 2 is practically inside the fault zone, it has different orientation of the anisotropy axis, corresponding exactly to the second reconstructed stress field. The possible interpretation is that it reflects superimposed deformations of the trend of Karlukovo Fault leading to its sinusoidal curvature.

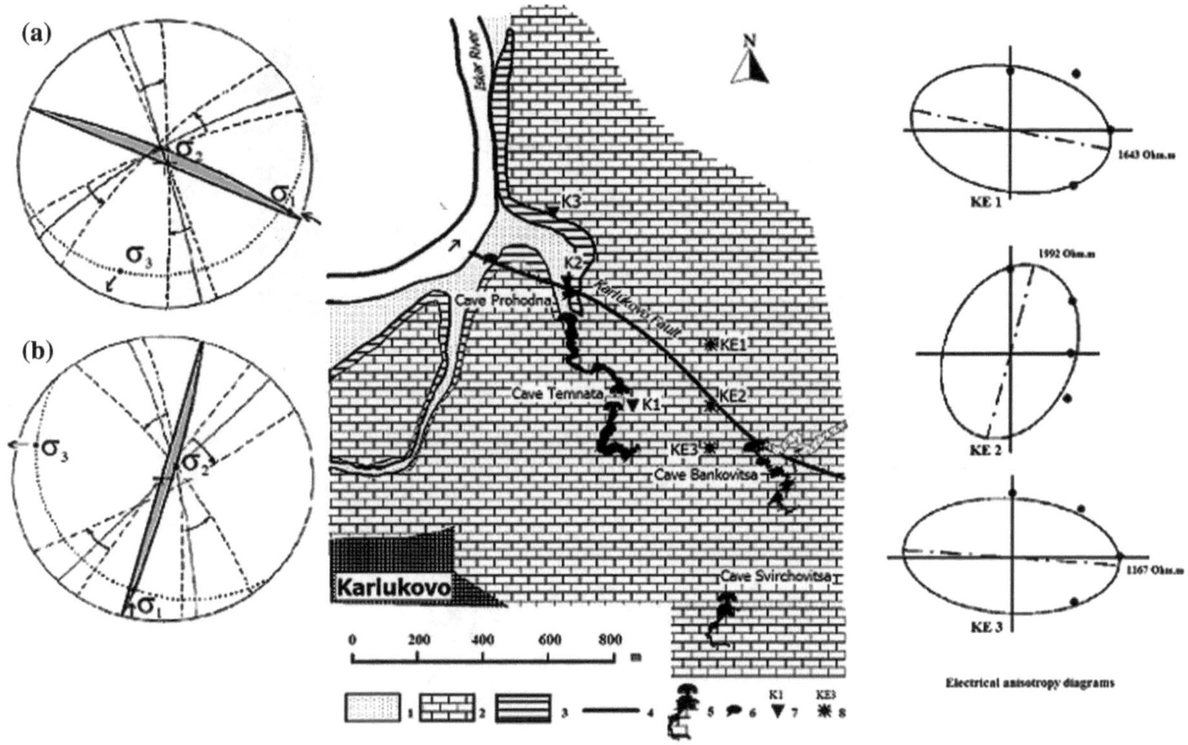

Fig. 2.42 Scheme of the studied territory NE of the village of Karlukovo with diagrams from the studied sites *1*—Quaternary deposits; *2*—Maastrichtian limestones; *3*—rock slope (canyon type); *4*—fault; *5*—cave entrance and horizontal projection of the galleries; *6*—karst springs; *7*—site of measuring of shear joints; *8*—site of study of the electrical anisotropy

Looking on the maps of the caves in the area (Fig. 2.42) it is clear, that the tendency of the cave galleries is to follow one of the preferential directions of open fractures, created by the principal tectonic impacts on the Maastrichtian limestones.

2.3.2.3 Intermediate Zone—Bosnek Karst Region

The karst system in Triassic limestones near the village of Bosnek (Southwestern Bulgaria) contains the longest Bulgarian cave - Duhlata (more than 17 km of total length of the galleries). Bosnek karst Region (BOS—Fig. 2.29), from tectonic point of view, belongs to Golo Burdo thrust unit (Jelev 1982; Gochev 1983; Zagorchev et al. 1994). Important structural and geophysical studies were done in this area for understanding the tectonic control on the karst formation and to explain better the hydrogeological features of the area (Benderev and Shanov, 1997; Shanov et al. 2001; Mihailova et al. 2006). Results from these studies will be analyzed and discussed below.

Geological Background

The karst-rock complex consists of Triassic carbonate rocks (Fig. 2.43). They overlie red sandstones, conglomerates and siltstones. The siltstones were transformed into quartzite at the contact zone of Vitosha pluton. The carbonate rocks cover an area of 23 km^2 and are about 200 m thick. Thick packages (about 50 m) of argillite and siltstone divide the carbonate rocks, playing the role of local aquicludes. The beds situated below these packages consist of limestone, and over them—of dolomite. Jurassic and Cretaceous rocks locally occur in the NE part of the area. A set of faults of different age of formation and activation has complicated the tectonic structure of the area, and in some cases, these faults control the relief, as well as the features of the karst system. The relief is typically mountainous—about 800 masl near the dam lake of Studena and up to 1,400 m in the NE part. The waters of Struma river principally, and also from other smaller rivers in the area, are the main factor for the karstification. The river loses its flow in the area of the contact between sandstone and carbonate rocks, as

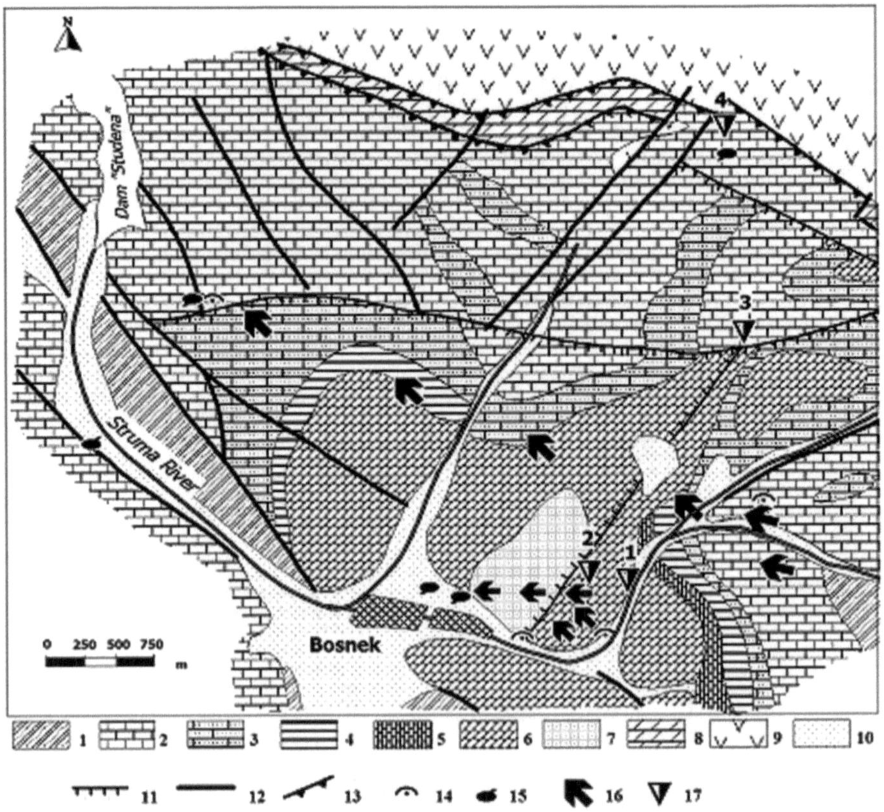

Fig. 2.43 Scheme of geological, tectonic and hydrogeological features of Bosnek Karst Region (after Benderev and Shanov 1997) *1*—Lower Triassic sandstones (T_1); *2*—Anisian limestones (T_2^a); *3*—Ladinian dolomites (T_2^l); *4*—Ladinian shales (T_2^l); *5*—Carnian limestones (T_3^k); *6*—Dolomites (Carnian-Norian—T_3^{k-n}); *7*—Limestones (Carnian-Norian—T_3^{k-n}); *8*—Siltstones and marls (Tithonian—J_3); *9*—Senonian volcanic rocks of Vitosha Mountain (K_2^s); *10*—Quaternary sediments; *11*—fault; *12*—normal fault; *13*—thrust; *14*—cave entrance; *15*—karst spring; *16*—principal direction of underground water flow; *17*—site of investigations of the rock fracturing

well as at the zones of the faults inter-crossing eastof the village of Bosnek. Two karst springs, situated in the northern part of Bosnek, disgorge their water again into the river of Struma (Fig. 2.43).

Exactly between the places of water losses and the springs near the village of Bosnek, the longest cave in Bulgaria named Duhlata is situated, with total galleries length over 17 km. Our studies on the tectonic stress fields and their impact on the hydrogeological peculiarities of the area (Benderev and Shanov 1997) have shown that tectonic stress fields created during the geological history systems of fractures, controlling the underground waters in the processes of karst formation.

The boundaries of Triassic carbonate complexes, are commonly tectonic that is predestined by the structural position of the discussed area. Orogenic processes and the related karst evolution are controlled by Pernik Fault Zone (Bonchev 1961, 1971; Kostadinov 1965; Matova and Angelova 1994) during Quaternary. Most of the faults crossing the area are practically segments of this zone with general strike NW–SE. Tectonic processes caused macro- and micro-block differential movements and fracturing of the rocks, leading to the formation of the contemporary valley of Struma River, as well as dividing the underground catchment, basins and formation of cave systems (Benderev and Shanov 1997). The processes of present day tectonic activity continue, and this is reflected by the swarm of seismic activity at 20 km northwestwards of the region during 2012 (Radulov et al. 2012), the strongest earthquake being of magnitude $M = 5.4$.

Karst Evolution

The investigated area was a part of Pernik basin during the Paleogene and was filled with sediments of

Middle Oligocene—Lower Miocene age. Probably, the Jurassic and Triassic sediments were overthrusted over Vitosha Pluton during Post-Oligocene time. Conditions were created for typical allogenous karst formation in a complex geologic-tectonic setting.

During the Pliocene-Pleistocene, as a result of the activation of orogenic processes, the morphological units around Bosnek Karst Region were shaped (Matova and Angelova 1994). The initial tectonic impulse was not so strong and for that reason the Struma River keeps its outflow direction toward west. Both the relicts of this old river valley and the accompanying them alluvial deposits could be seen SE of Bosnek Village.

The Quaternary stage begins with a change in the intensity of tectonic processes. As a result of repeated movements and subsequent vertical tectonic deformations, the rock complexes affected by karstification were exposed on the surface and the formation of an underground karst complex started. The valley and karst complex development are considered as related to the changes of the character of the tectonic movements, phases of the glacial epoch, lithostructural features and Quaternary tectonic activity. During the Upper Pleistocene, segments of Pernik Fault Zone were activated. The NW-SE trending faults of Bosnek Karst region belong to this fault zone. As a result of this process, two separate areas are formed in Bosnek Karst Region. The tendency of regional uplift and river incision continued after the Middle Pleistocene. The karst processes from the Pleistocene up to now are determined by Struma River, being simultaneously a rivershed and a drainage system (Benderev and Angelova 1999). The faster block uplifting of Vitosha Mountain led to the formation of a steep slope of Struma River valley and consequently the river water begun to flow underground in the zones of carbonate rocks.

The alternation of active uplift and relatively low tectonic activity led to formation of different levels of galleries, some of them even under the contemporary river valley. As a fact proving the presence of different stages in the karst development, mainly the structure of Duhlata Cave system should be considered where the cave galleries are much longer and where seven horizontal levels could be clearly delimited. These levels and their number correlate closely with river terrace fragments found on the valley slopes (Benderev and Angelova 1999). The oldest and hypsometrically highest galleries could be considered as the most upper and oldest way of underground water flow that corresponds to the planation during the Pliocene-Pleistocene time. Thus, the age of that process could be accepted as a beginning of a more intensive karstification of the whole Bosnek Karst Region. As a result, a complex multi-leveled labyrinth cave with a basic underground stream and a few small tributaries has been created. The process of the cave morphogenesis continues up to present days.

Tectonic Stress Fields Reconstructions

The reconstructions of the tectonic stress fields were made using the statistical approach for analyzing the conjugated shear joint systems, the kinematics of the deformations along faults discovered in Duhlata Cave (well-preserved striations on slickensides), and the fault-plane solutions for some local earthquakes. The studied sites are outcrops of limestones, as well as inside the Duhlata Cave (Fig. 2.44). Initially, two expressed tectonic stress fields have been reconstructed (Table 2.6).

The older stress field is determined to be from Post-Oligocene time (Benderev and Shanov 1997). This stress field reflects the tectonic phase of trusting of the Mesozoic rock complexes over the volcanic rocks of Cretaceous age of Vitosha Pluton. The studies in the cave discovered one older tectonic stress field. The orientation of the minimum tectonic stress axis σ_3 is subvertical whereas the maximum tectonic stress axis is sub-horizontal, and this is a clear indication for trusting regime of the deformations.

The inferred second tectonic stress field is of Pliocene-Quaternary age. The tectonic compression was directed NW–SE and the active sub-horizontal NE–SW tectonic extension created possibilities for opening of the NW–SE fractures and their karstification. The logical consequence was that the karst galleries and the way of the underground waters have the same general direction (Fig. 2.43). Namely this deformation is reflected in the movements along fault No. 2 in Fig. 2.44 and the formation of slickensides on the fault surface.

During the works in the cave Duhlata we discovered also the second generation of tectonic striations. They are well preserved in the cave and are probably very young because the clay is remaining on the striations (Fig. 2.45). The solution practically confirms the Pliocene-Quaternary orientation of the

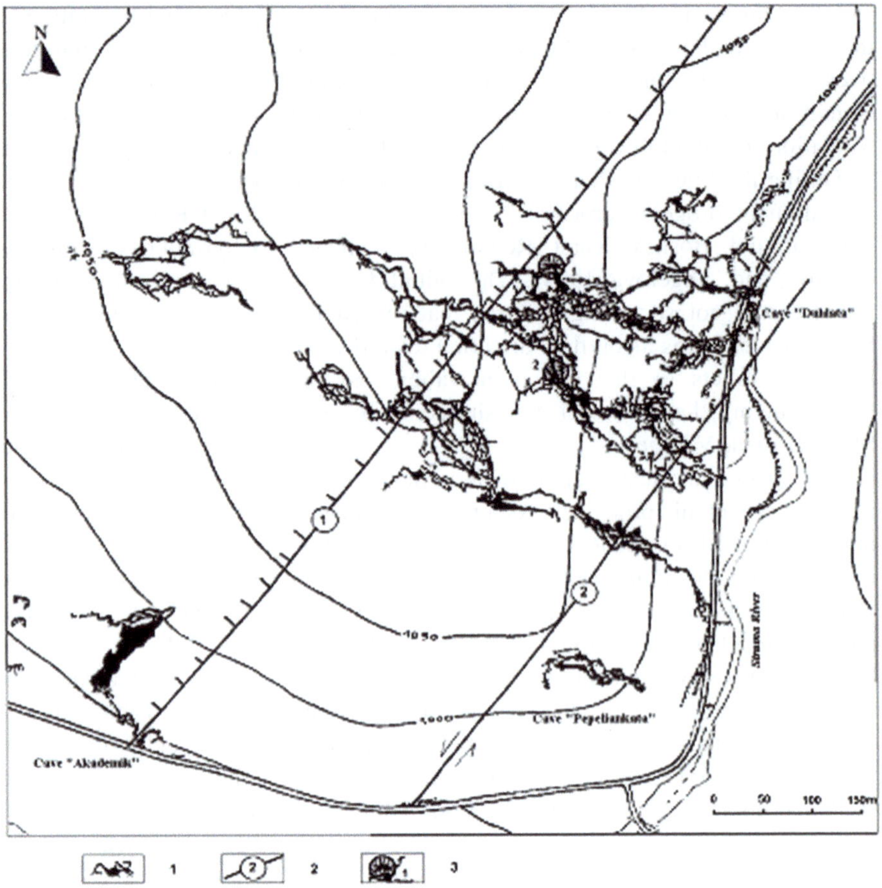

Fig. 2.44 Scheme of the caves in Bosnek Karst Region in the vicinity of Cave Duhlata (based on the scheme from Benderev and Angelova 1999) *1—horizontal projections* of the karst galleries; *2*—fault; *3*—site of investigations inside the Cave Duhlata, Bosnek Karst Region, Bulgaria

Fig. 2.45 Preserved tectonic striations in Cave Duhlata, Bosnek Karst Region, Bulgaria

Table 2.6 Reconstruction of the sequence of tectonic impact on the rocks of Bosnek Karst Region

Origin	Site №	Post-Oligocene	Pliocene-Quaternary	Contemporary	Method
Earthquake 22.05.2012 $M_w = 5.6$ N 43.65^0 E 23.00^0					Fault-plane solution
T^{k-n}_3 **dolomites**	1 and 2		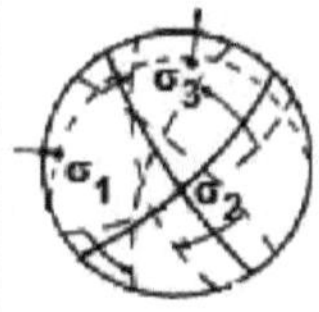		Shear joints
T^{k-n}_3 **dolomites**	1 and 2	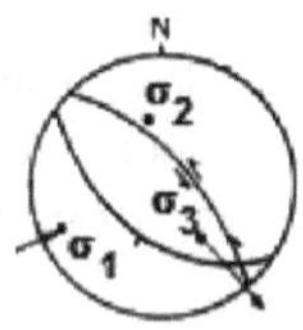	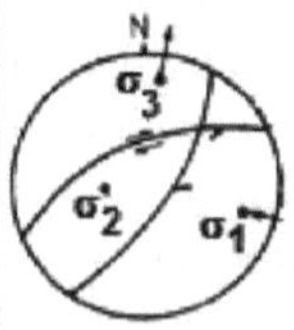		Striations on slickensides
T^a_2 **limestones**	3				Shear joints
T^a_2 **limestones**	4		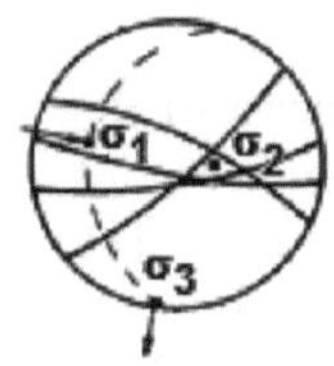		Shear joints and striations on slickensides

tectonic stress axes—sub-horizontal E–W compression and N–S extension.

Due to the fact that some of the older karst forms were cut and displaced, and because of the position of the studied area between two very active seismic zones in this part of the Balkans, we think that the detected traces of young movements can be related to a past seismic event (or events). The nearest reconstructed fault-plane solution from 2012 Pernik Earthquake is supporting the opening of the NW–SE oriented fractures (Table 2.6), which is the general trend of the karst galleries of the caves.

The age of the left-lateral movement along fault No. 2 (Fig. 2.44) is not clear. This fault was known from the geological mapping in scale 1:25,000 only northeastwards from the karstified area. It displaces sediments of Jurassic age. We discovered its extension toward SW and a slickenside with striations, showing also left-lateral displacement. The entrance of Cave Duhlata lies exactly on this fault. The fault is expressed in the relief, it stops the galleries of Pepeliankata Cave, but its impact on the karst systems is not very well expressed.

Clearer control on the karst processes can be found along fault No. 1 (Fig. 2.44). Namely, a subsidence of the NW block is noted in the area inside the cave of Duhlata. In this block the density of the karst galleries is lower and the mapped galleries are active. This fault is also controlling the morphology of the Akademik Cave (Fig. 2.44). One possible reason for the expressed impact of this fault on the karst system could be paleo-earthquake activity. The sudden subsidence of the NW block adjacent to the fault could juxtapose its old galleries to the active ones of the SE block. The younger galleries of the NW block have been filled with clay or other sediments, or are inside the phreatic zone. This assumption is supported by the fresh striations on the fault surface in the cave of Duhlata. There are a lot of broken or inclined stalagmites, but the caves of the region are frequently visited by a number of people and thorough studies have to be made for filtering the naturally from the artificially broken stalagmites.

2.3.3 Cuba: Structural and Geophysical Study of the Karst System of Guaso Plateau (Eastern Cuba)

2.3.3.1 Regional Characteristics

Guaso Plateau is situated 1,000 km southeast of the City of Havana (Republic of Cuba), in the district of Guantanamo, Province Sierra Oriente (Fig. 2.46). The studied region by the Bulgarian-Cuban expedition in 1988 covered a territory of 230 km^2, occupied by tropical forests.

The altitude is between 100 m and 800 m above sea level and the plateau is cut by canyon-like, narrow valleys. This difference in the altitudes creates a situation, where the average annual temperature changes from 20 to 23 °C in the upper part of the plateau to 24–27 °C at its foot. The plateau is a barrier for the humid air from the North, from the Atlantic Ocean. The abundant rains, even during the dry season, form significant surface and underground water flows toward North and only a part of the water runs toward South as underground current inside the 500 m thick layer of Eocene limestones (Formation Charko Redondo). Due to the geographical location of the plateau in the subtropical zone, a seasonal dynamics of the rainfall exists, with 70–76 % of the rains during the humid season (May–October). The water balance, calculated by Cuban hydrogeologists has proved that considerable part of the water is drained to the South, toward the Caribbean Sea, using unknown underground ways. This water is extremely needed for the normal water supply and development of the town of Guantanamo and its vicinities. The main problem of the expedition was the detection of the underground way of the water and to improve the possibility to exploit sufficient quantity of potable water from the karst system. The base for the studies were the topographic maps in scale 1:50,000, the geological map in scale 1:50,000 drown by Hungarian and Cuban geologists, as well as the aerial photos of the region. The results were only partially published (Andreychuk and Benderev 1991; Chanov and Benderev, 1993).

2.3.3.2 Geological and Hydrogeological Background

The rock massif of Guaso Plateau is built mainly of carbonate rocks. The geological sketch of the studied area is shown in Fig. 2.47. The following stratigraphic units are present:

- BUCUEY (BUC)—K_{1-2}^{al-t}—magmatic, metamorphic and volcanic rocks with total thickness more than 500 m. The outcrops are located predominantly in the northern part of the plateau. On some of these sites, a thick weathered soil *(terra rossa)* was formed, containing manganese-iron concretions. On the upper part of the plateau, over the sedimentary rocks, remnants of an allochtonous serpentinite overthrust can be found.
- PICOTA (PIC)—K_2^{cp-m}—conglomerates and breccias with thickness varying between 100 and 150 m.
- CHARCO REDONDO (CHR)—P_2^2—Eocene limestones with thickness from 200 to 500 m. In these rocks intensive processes of karstification formed numerous caves, hollows, abysses, underground rivers and lakes. This formation was the principal target for exploration from the Bulgarian-Cuban expedition during 2 months. The total length of the mapped karst galleries was about 20 km, 8,382 m of them being active galleries with underground river in the longest cave El Campanario. Unfortunately, in

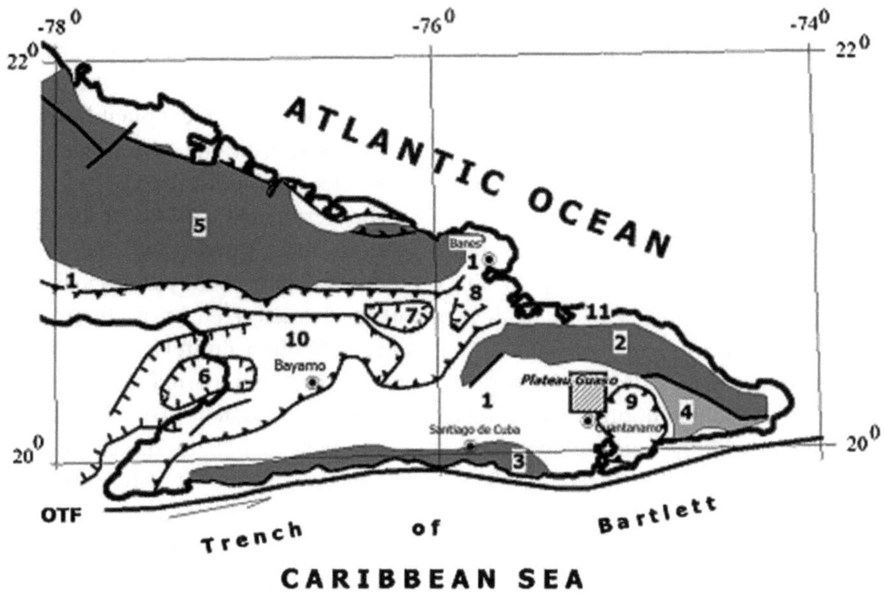

Fig. 2.46 Schema of the principal tectonic units of Eastern Cuba. The studied territory is hatched. *OTF*—North Caribbean (Oriente) Transform Fault *1*—Synclinorium of East Cuba; *2*—Anticlinorium Mayeri-Baracoa; *3*—Anticlinorium of Sierra Maestra; *4*—Horst structure of Sierra del Purlal; *5*—Anticlinorium of Camagüey; *6*—Depression of Guacanayabo; *7*—Depression of Cacocum; *8*—Depression of Nippe; *9*—Depression of Guantanamo; *10*—Depression of Cauto; *11*—Depression of Bahia de Nippe

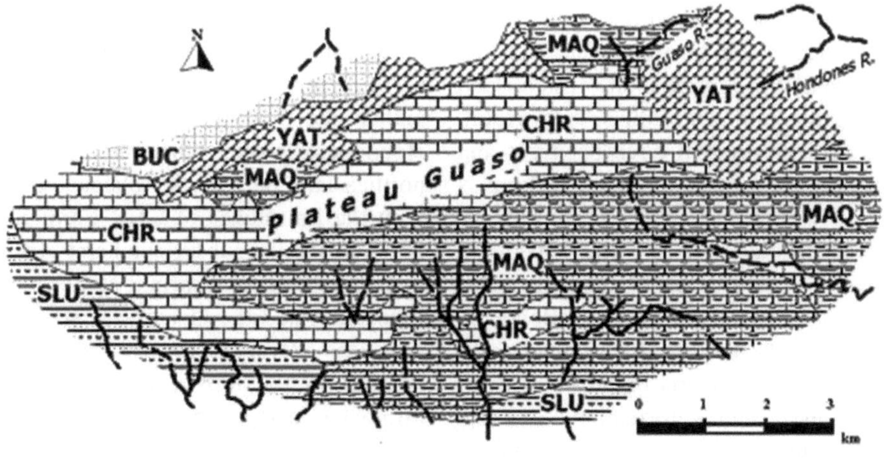

Fig. 2.47 Generalized scheme of lithostratigraphic units in the vicinity of Guaso Plateau (Eastern Cuba) and main temporary and permanent rivers. *Formations*: BUCUEY (*BUC*)—K_{1-2}^{al-t}; CHARCO REDONDO (*CHR*)—P_2^2; SAN LUIS (*SLU*)—P_2^{2-3}; MAQUEY (*MAQ*)—P_3^3—N_1^1; YATERAS (*YAT*)—N_1^{1-2}

the last day of the expedition, all drown maps were confiscated by Cuban military representatives.

- SAN LUIS (SLU)—P_2^{2-3}—Sedimentary formation of clays, marls and thin clayey sand layers. The total thickness of this formation is about 350 m. It covers the plane southwards from Plateau Guaso, overlying the limestones of Charco Redondo Formation.
- MAQUEY (MAQ)—P_3^3—N_1^1—Molasse formation represented by 200 m thick beds of marls, clayey limestones or marls containing thin limestone layers. Everywhere, on the top of the plateau, it covers the limestones of Charco Redondo Formation and its sediments fill up the negative forms of the paleo-relief.

- YATERAS (YAT)—N_1^{1-2}—Organogenic, coral-reef limestones with a thickness not exceeding 50 m. Exotic blocks of this formation can be found in the sediments of Maquey Formation. It is supposed that some surface rivers at the top of Guaso Plateau have lost their waters when reaching the rocks of YAT Formation.
- Quaternary (Q)—Soils with thickness from some centimeters to several meters. At the top of the plateau, there are negative forms with 25 m thick clay layer with rock debris.

The tectonic background (Fig. 2.46) is assumed to be a result of the geodynamic evolution of the Cuban micro-plate at the contact with the oceanic plate—the trench of Bartlett (Lilienberg 1983; Cobeilla 1984; Rojas-Agramonte et al. 2003). This trench predetermines the existence of Oriente fault system (N 80–90°) with expressed sinistral strike-slip movement. It has affected the southern Sierra Maestra mountain range, SE Cuba, and portrays the dynamics and tectonic evolution of the southern Cuban coast in the boundary zone between the Caribbean and North American plates (Rojas-Agramonte et al. 2003). The region has been affected by historical and recent earthquakes. The fault-plane solutions from earthquakes along the trench are in agreement with the dominant ENE-WSW compression established onshore (Rojas-Agramonte et al. 2003). As a result, the faults of the Oriente system have been cut and displaced by a younger system of faults striking N 40–60°.

The hydrometrical studies during the expedition have shown that 1.2 m^3/s of water is drained by the karst system of Guaso Plateau (Chanov and Benderev 1993). The feeding of the karst waters is from infiltration of surface waters, especially in the areas where Charco Redondo and Yateras formations outcrop. Guaso River disappears in a sink-hole near the cave of Sumitero with a discharge of 300 dm^3/s (middle of January, 1988), and reappears as a river from the cave El Campanario with a discharge of 500 dm^3/s (Andreychuk and Benderev 1991). The modulus of the underground stream is approximately 8 $dm^3/s\ km^2$. Taking into account the season of the recording, the quantity is the minimal one. After rains the quantity of the water rises considerably because the high permeability of the karst system. As an example, the discharge rate of the river from the cave El Campanario raised to 15–20 m^3/s during 36 h after strong rainfall in January, 1988. Practically, at that time, these considerable quantities of water were not exploited, except these from the Spring Bataldo, used for water supply of local plant for production of sugar, and the river from the cave El Campanario, used by a hydro-electrical power plant.

The performed geological and geophysical investigations have proved that the karst system can be subdivided into two hydrological units. The western part is characterized by a clearly expressed zone of water saturation entirely inside the karst cavities of Charco Redondo Formation. The reason of its formation is the existence of a fault barrier at the foot of Guaso Plateau, where the karstified rocks of CHR contact the clays of SLU Formation. Along this E-W trending border spillways exist, being at the same altitude and with the same chemical composition of the water. According to geophysical data, a karstified layer can be expected southwards of this line, representing a confined aquifer.

The geological conditions in the eastern part of Guaso Plateau are more complicated. MAC Formation is of wide occurrence and is resistant to karstification. The existing large karst negative forms on the surface are filled up with *terra rossa*. These traits and the tectonic faults contribute to the formation of local barriers for the water. Two rivers on the surface gather the waters from the plateau—Guaso and Hondoness Rivers. Hondoness River appears in the northeastern part of the plateau, but at the contact with the limestones of CHR Formation it is flowing down inside the subsurface karst system. Downwards it appears as several springs at the contact between the limestones and the Quaternary sediments.

Guaso River is formed by a number of karst springs on the top of the plateau and after some kilometers disappears in Somitero of Guaso, forming a spectacular underground river, successfully traced by the speleological studies in the caves El Campanario and Somitero. The karst system there has considerable water permeability—more than 100 m^3/s.

The performed geophysical works in 1988, using electrical methods (vertical and azimuthal electrical soundings), permitted detection of karst forms with considerable dimensions, discovered and mapped later by the speleologists. The dimensions of the galleries are 20 m and more in diameter and their tops are only 10–20 m below the surface. These significant dimensions of karst galleries, formed inside relatively

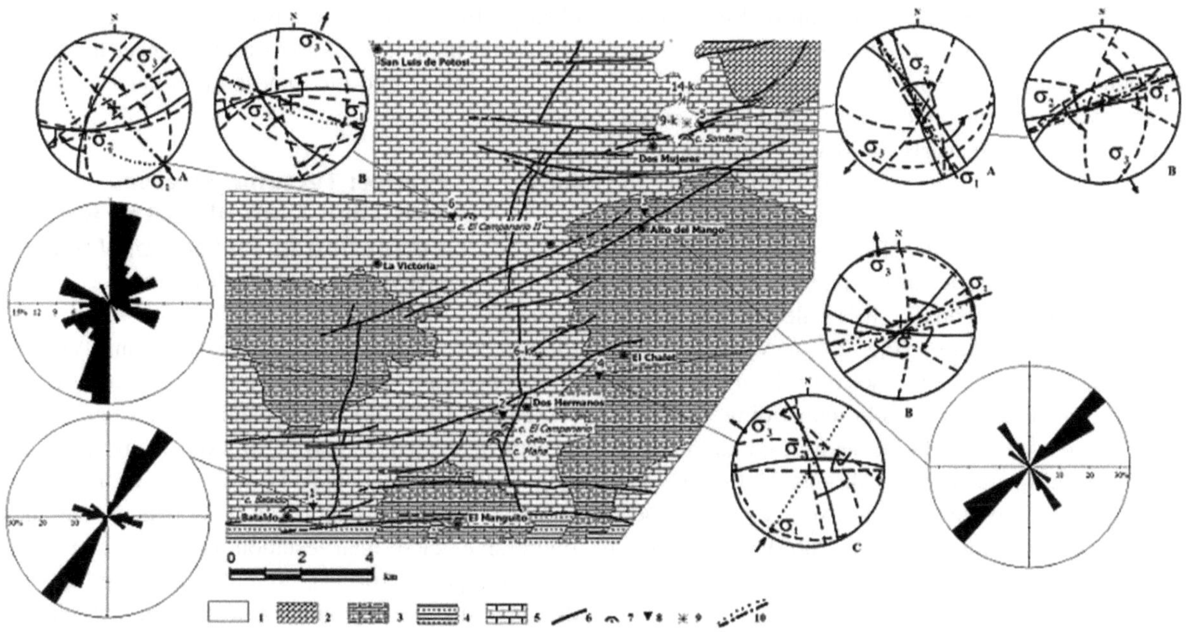

Fig. 2.48 Detailed geological map of the studied territory on Plateau Guaso (Eastern Cuba) and reconstructions of the tectonic stress fields based on measurements of joints (on the base of the map elaborated by K. Bonev) *1*—Quaternary sediments—*Q*; *2*—Formation YAT; *3*—Formation MAQ; *4*—Formation SLU; *5*—Formation CHR; *6*—fault; *7*—cave entrance; *8*—site of structural measurements; *9*—site of studies of the electrical anisotropy; *10*—most probable position of the open fractures on the stereograms

young rocks, are due to the tropical conditions of karst formation and evolution.

2.3.3.3 Tectonic Stress Fields and Their Control on the Karst System

Two methods were applied for the analysis of the tectonic stress fields—Nikolaev's Method (Nikolaev 1977), based on the dispersion of the shear joints density and studies of the electrical anisotropy. Additionally, at three sites only, the directions of few fractures have been measured (number from 10 to 20) and rose-diagrams were plotted in Fig. 2.48—sites 1, 2 and 3. These rose-diagrams reflect normally the strike of the adjacent lineaments and they are not sufficiently informative for reliable geological conclusions.

Quite different, as quality, is the information from the sites of mass measurements of shear joints. It permitted the reconstruction of the tectonic stress fields and the analysis concerning their impact on the karst formation and evolution (Fig. 2.48—sites 4, 5 and 6). The tectonic interpretation of the diagrams of the electrical anisotropy has been made also in the context of the tectonic stress field reconstructions and the scheme of the fault systems drown on the base of remote photos and field verifications (Fig. 2.48).

Sites 5 and 6 characterize fracturing of the limestones of CHR Formation. Two tectonic stress fields are detected in this formation. The first one (*A*) and maybe the oldest for this formation has provoked the opening of fractures with NW–SE direction. The second one (*B*), being of great importance for the recent tectonic situation in the region, determines the existence of sub-equatorial fractures and faults, controlling the step-like subsidence of CHR Formation to the south. This tectonic stress field acted after the consolidation of the sedimentation of MAQ Formation, because it was also reconstructed using the joint set in the marls of this formation (Site 4). The variations in the directions from N 80° to 110° correlate well with the regional faults near the sites of measurement. The sub-equatorial faults are in some places barrier for the karst processes and re-distributor of the water. The commented tectonic stress field coincides with deformation phase *D*1 of N–S extension, exemplified mainly by karst-filled extensional veins and normal faults (Rojas-Agramonte et al. 2003). These authors correlated this deformation with the regional kinematics in the Caribbean realm, imposed by the opening of the Cayman trough and the disruption between Cuba and Hispaniola.

The youngest tectonic stress field was reconstructed only in the marls of MAQ Formation (Site 4—*C*). This tectonic stress field determines the opening of a sub-vertical system of fractures with general direction N 30°. This is also the general direction of the galleries of the largest cave system in the region—El Campanario. It is reflected also by the rose-diagram from Site 3 (Fig. 2.48). At the entrance of El Campanario Cave the rose-diagram of the fracture orientation is mostly coinciding with the strike of the system of faults with orientation favorable for opening under the conditions of stress field C. A number of cave entrances are situated along these faults. Rojas-Agramonte et al. (2003) reconstructed deformation phases *D*2 and *D*3 comprising NE–SW to nearly N–S directed compression.

Sites 6-k, 9-k and 14-k in Fig. 2.48 are the places of azimuthal vertical electrical sounding. The corresponding graphs are plotted on Table 2.7.

The graphs show the general tendencies of the electrical anisotropy on the site of measurements. Sites 6-k and 14-k are informative for the limestones of CHR Formation, while site 9-k is on a large negative karst structure with a tick filling of young, mostly Quaternary sediments.

The ellipses of anisotropy at point 6-k are for depths not exceeding 10 m and their general direction of the long axes N 95° reflects the controlling role on the fracturing of the nearest fault with the same direction. But the site was situated exactly over the known galleries of Campanario Cave and the abrupt changes of the direction of the anisotropy ellipse to azimuths between N 5° and 50° is reflecting the preferential direction of the cave galleries. The situation is similar for point 9-k where a clear coincidence exists between the long axes of the electrical anisotropy ellipses (N 50°) and the detected lineaments with SW-NE direction. The young age of this direction was discussed above and probably it controls the recent karstification. It can be also assumed that this is the direction of a satellite fault to the well expressed regional fault zone of E–W direction.

The E-W direction is well reflected in the ellipses of anisotropy from point 14-k. Even to depth more than 100 m, the direction N 90–95° is so stable that it clearly illustrates the controlling role of the sub-equatorial fault on the karst structures near the falling of Guaso River into the underground karst system.

2.3.4 France: Tectonic Stresses and Their Control on the Karst Formation in Plateau of Vaucluse (Alps Maritimes, France)

Plateau of Vaucluse, framed by Mont Ventoux (1,912 m) from northwest, by the Mountain of Lure (1,826 m) from northeast, and by the geological structure, named Syncline of Apt from the south (Fig. 2.49), is worldwide known with its karst phenomena. Spring Fontaine de Vaucluse is the most impressive karst manifestation in this region (vauclusian type of spring). The 308 m deep abyss is draining the underground waters from the plateau. The coloring test of waters from different places on the plateau or in the caves has shown a relation between the caves and the Spring of Vaucluse (Mudry and Puig 1991). Among a number of publications concerning the karst of this area, a study on the relationship between karst formation and recent tectonics in the context of the tectonic stress fields has been published (Shanov and Cousset 1993). More detailed analyses, including reference on geodetic records, completed these studies (Shanov and Georgiev 2001). The presented analyses follow the principal results from these two publications.

Southeastern France is also an area of earthquake activity and this fact allows for a more complete evaluation of the contemporary tectonic stress field and the related processes in the karst.

2.3.4.1 Geological Background

Plateau of Vaucluse is formed by limestones and marl-limestones of Lower Cretaceous-Hauterivian, Barremian and Bedoulian age, as well as by rocks of the Urgonian reef facies (after Fage 1981). The Provencal tectonic phase at the end of the Eocene has marked the uplift of Plateau of Vaucluse and the beginning of the karst processes in the limestones. The contemporary morphological features were formed during the Post-Pontian tectonic phase, expressed by the overthrusts of Mont Ventoux and of Lure Mountain, when the Provencal tectonic block

Table 2.7 Graphs from the azimuthal studies of the parameters of the electrical resistivity on Plateau Guaso (Cuba)

Site №	Azimuth [N ^{0}]	Electrical anisotropy	Eccentricity	Electrical resistivity [Ohm.m]
6-k	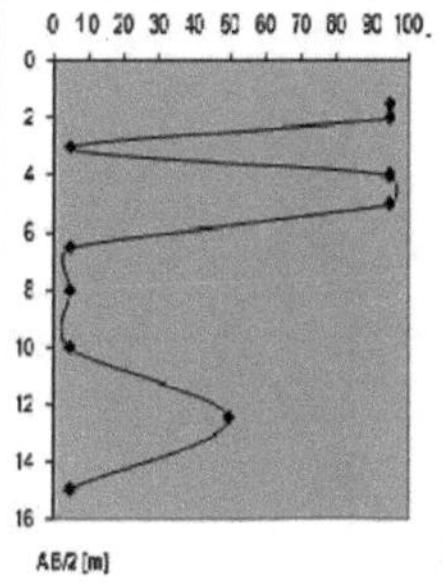	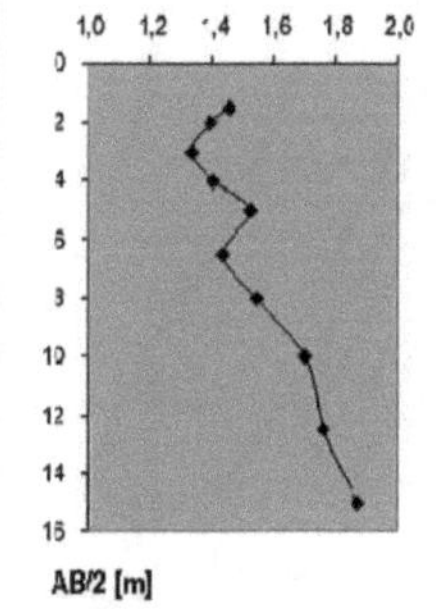	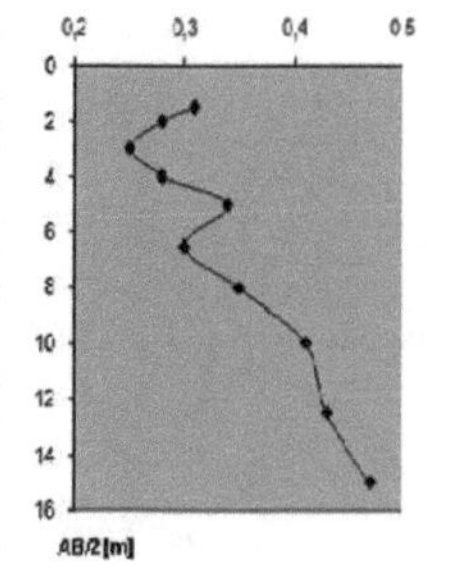	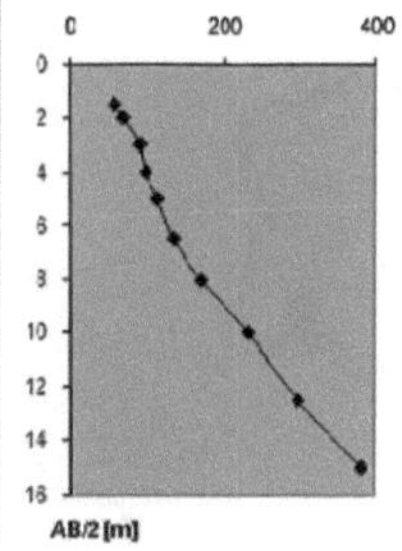
9-k	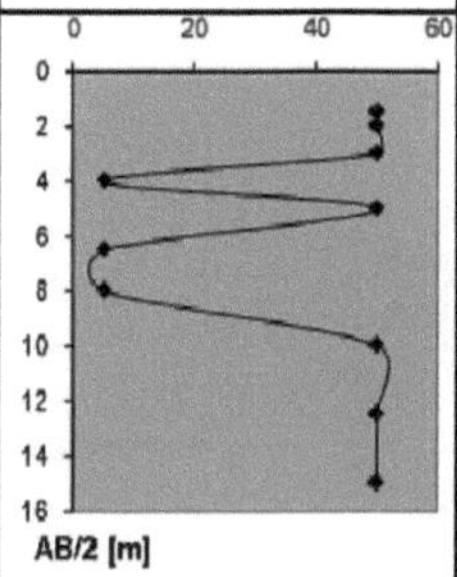	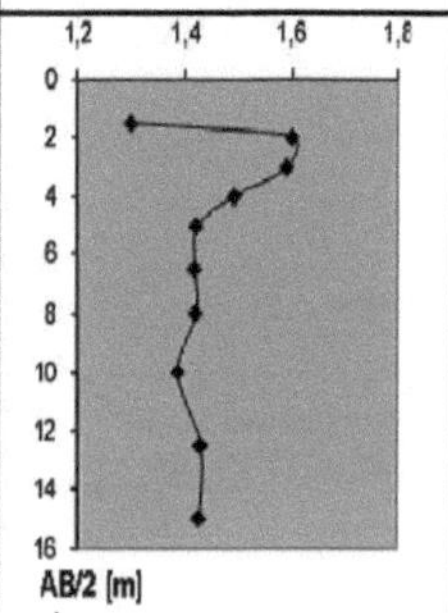	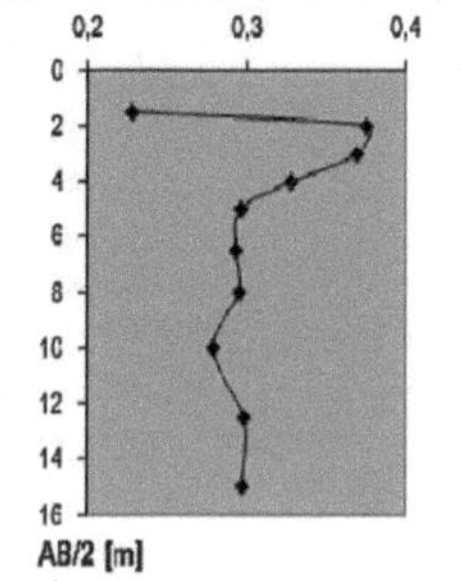	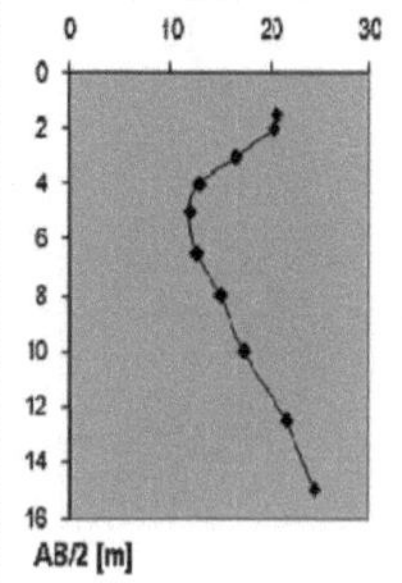
14-k	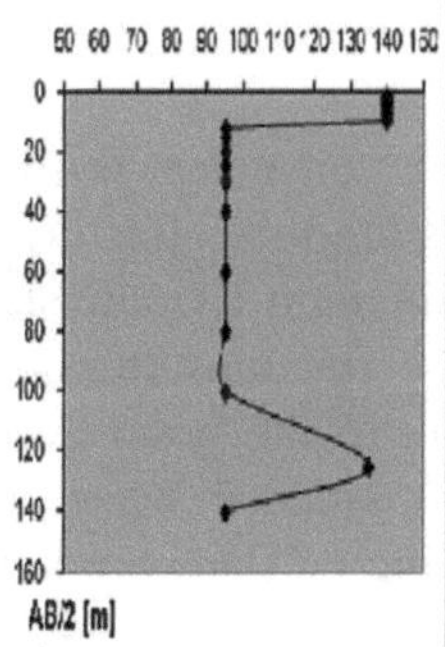	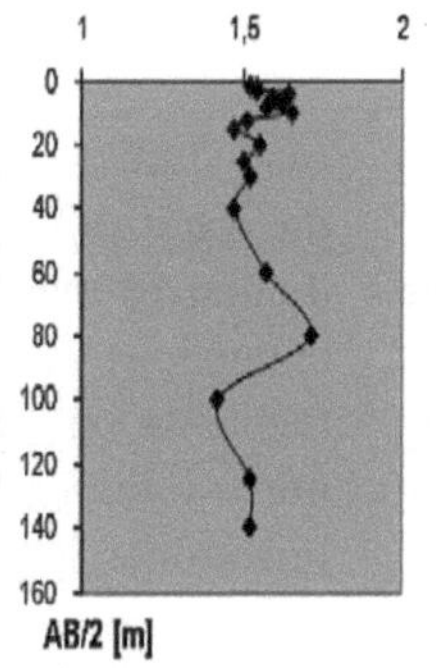	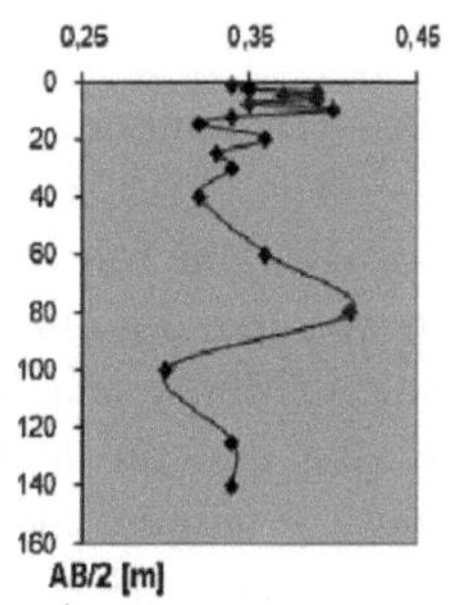	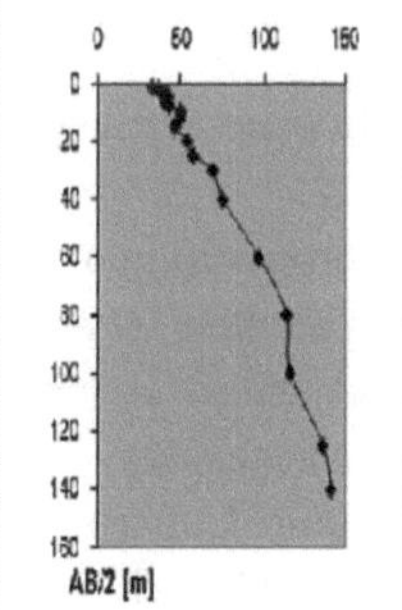

moved toward north. During this period, the NW–SE strike-slip faults formed (Fig. 2.49). The better expressed Post-Burdiganian (end of Oligocene) phase should be noted, when NE–SW striking faults were formed, controlling to a great extend the recent karst evolution in the rock massif.

Analyzing the genesis of Plateau of Vaucluse on the base of the principal forms of surface karst, Depambour and Guedon (2004) discovered recently deformed paleo-surfaces, but they have originated on a regional scale since Post-Miocene time. All intermediate surfaces (alt. 850 m) cut the Miocene deposits (Burdigalien) of Aurel—Sault Depression. This reveals a major phase of Post-Miocene denudation. These surfaces of karstic flattening are assumed to be related to the vast poljes of the sectors of Saint Christol and Sault. Their genesis, dependent on dysfunctions of the endokarstic drainage, could be related

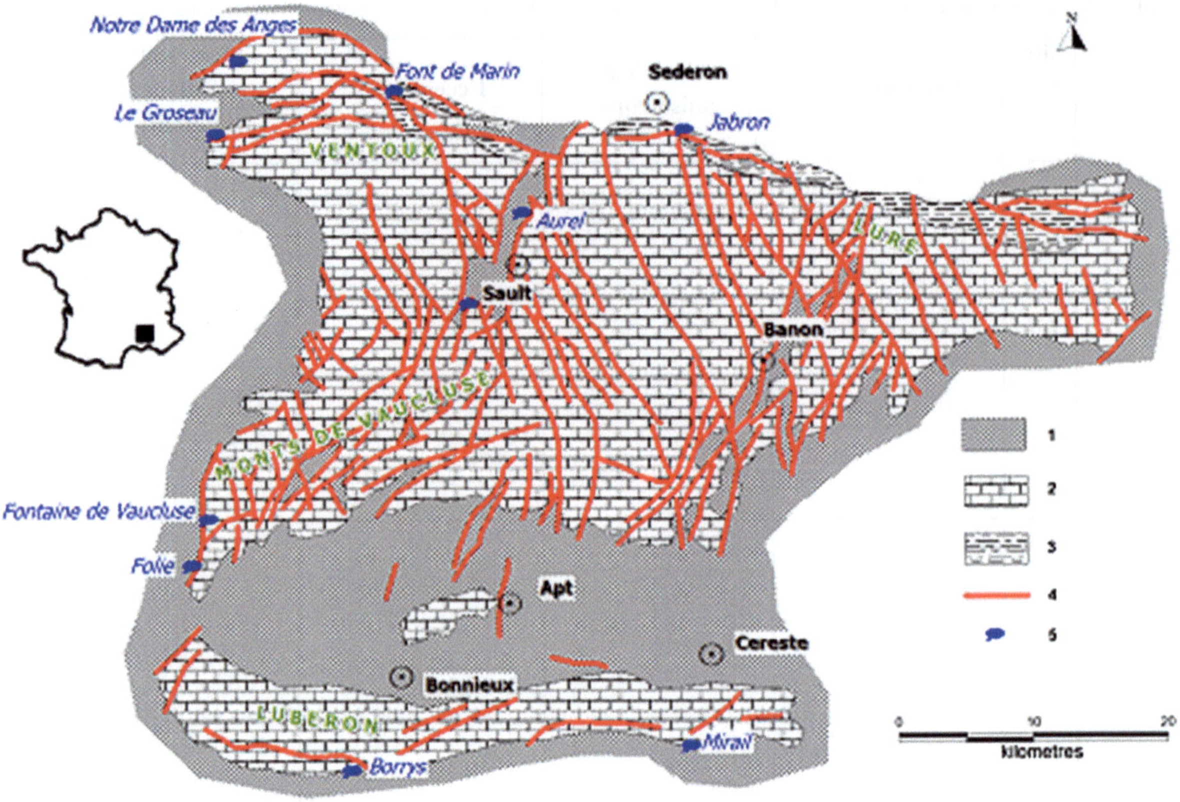

Fig. 2.49 Geological scheme of the region of Vaucluse (after Mudry and Puig 1991) *1*—Bedoulien terrain; *2*—Lower Cretaceous limestones; *3*—Neocomian marls; *4*—fault; 5—spring

to the Pliocene transgression (Depambour and Guedon 2004).

From hydrogeological point of view, the role of the spring of Vaucluse (Fig. 2.50) for the draining of underground waters is well studied (Murdy and Puig 1991). It is necessary to note also that the hydrogeological unit Luberon-Vaucluse continues below the syncline of Cavaillon (Gouvernet et al. 1971) and this was confirmed by the borehole near Chene, 5 km westwards from the town of Apt.

2.3.4.2 Methods of Study

Fracturing

Mass measurements of the space elements of joints (dip direction and dip angle) in different outcrops of the rock massif were done and they provide the basic data for applying the method of Nikolaev (1977) for reconstruction of the main axes of the tectonic stress fields. As shown above, when presenting the different case studies, most convenient for the development of the karst process are these fractures that are perpendicular to the minimum tectonic stress axis (tectonic extension). These fractures (tensional fractures) are the least resistant to the water current and they are most effectively widened when dissolution of the limestones is going on (Dreybrodt 1988).

Karst Galleries Orientation

The diagrams of the orientation of the karst galleries could be useful for indication of the main preferred ways of the underground waters. Mostly, this is the direction of the open fractures and the practice has shown that this type of fractures in the recent active karst systems is normally connected with the youngest tectonic stress field.

Earthquakes

The most appropriate method for determining the main directions of the contemporary tectonic stress

Fig. 2.50 Principal hole of Spring of Vaucluse in summer season (low water level)

field are the fault-plane solutions from earthquakes (ex.: Mattauer and Mercier 1980). Every fault-plane solution for a given earthquake focus is closely related to the local tectonic conditions. They often show a remarkable coherence for large areas. The concentration or the alignment of epicenters of weak earthquakes in some territories helps to detect the contemporary active tectonic zones.

Satellite Laser Ranging Geodetic Solutions

The Grasse laser tracking station velocity (SE France) is derived from the analysis of 977,000 normal points of the geodynamic satellite Lageos-1 covered the data span of 16 years—from January 1984 till December 1999 (Shanov and Georgiev 2001). The data reduction and analysis procedure were performed by the SLRP 3.2 software for dynamic orbital analysis and least-square parameter estimation (Georgiev and Kotzev 1995) using the recent dynamicmodels and fundamental constants (McCarthy 1996).

2.3.4.3 Tectonic Stress Fields and Analyses of Their Control on the Karst System

Data on shear joints analysis from the limestones of the plateau have been collected from five outcrops (Fig. 2.51). The elements of 100 joints at least were measured on every site. After mathematical processing and the corresponding interpretation, two tectonic stress fields have been reconstructed, both having remarkable coherence in directions of maximum and minimum tectonic stress axes for the different sites. The characteristic feature of both stress fields is the sub-horizontal position of the maximum and minimum stress axes. The problem was to determine the relative age of every one of the tectonic stress fields.

The elaboration of rose-diagrams, indicating the preferential directions of the horizontal galleries, has been made using the maps of the cave galleries in the region (Fage, 1981). It is evident (Fig. 2.51) that, in general, the NE–SW direction is preferential and there is a close relationship between the open fractures, indicated on Table 2.8 as product of the Post-Burdiganian Phase, and the karst galleries. This direction is also well marked by the general strike of the youngest faults.

The scheme of these faults is elaborated on the base of two principal sources of information—The Map of Remote Sensing Lineaments of France in a scale 1:1 000 000 (Scanvic and Weecksteen 1984) and the Seismotectonic Map of France (Voght and Godefroy 1981). The lineaments from satellite images could be accepted under some conditions as traces of reactivated faults or faults formed during the neotectonic stage and active up to now. The faults from the Seismotectonic Map are interpreted as structures,

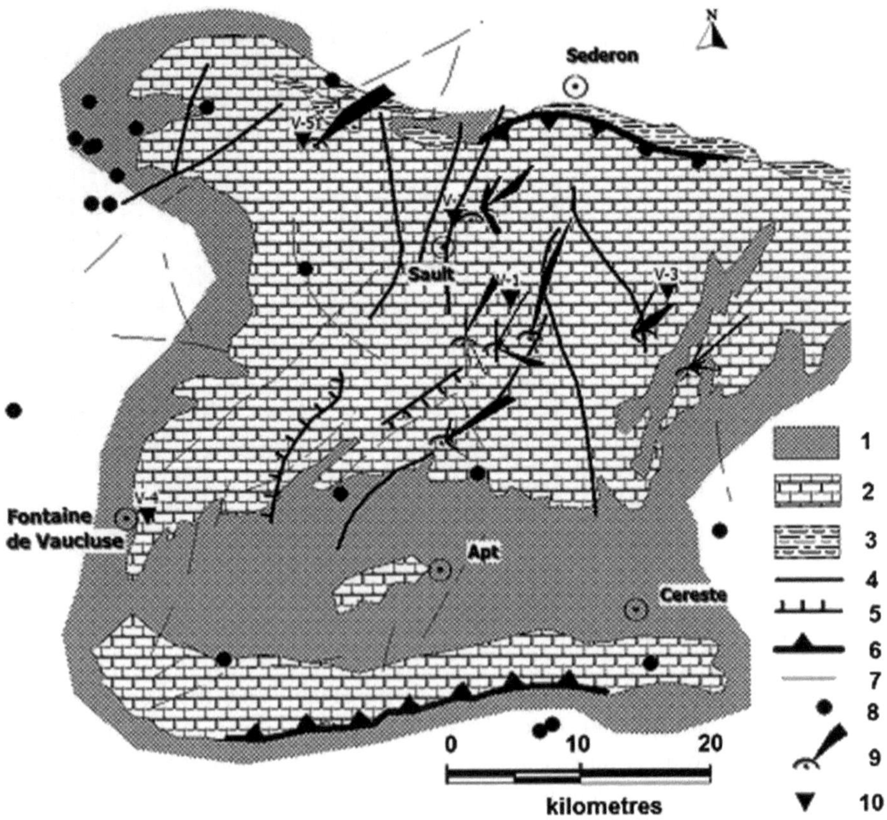

Fig. 2.51 Geological scheme of Plateau of Vaucluse with elements of geodynamics (after Shanov and Cousset 1993; Shanov and Georgiev 2001) *1*—Bedoulien terrain; *2*—Lower Cretaceous limestones; *3*—Neocomian marls; *4*—fault; *5*—normal fault; *6*—front of thrusting; *7*—remote sensing lineaments; *8*—earthquake epicenters (magnitudes $M_L = 1.0–3.0$ and time interval from 1982 to September 2013); *9*—cave location with rose-diagram of the karst galleries; *10*—point of mass measurements of joints

connected directly to the present day tectonic processes. They could be distributors of the seismic energy or generators of this energy. For the monts of Vaucluse and Ventoux a clear coincidence exists between the main strike of these faults (NE–SW) and the preferred direction of karst processes, in spite of the well-developed fault system with general trend NW–SE (see Fig. 2.49). Such a result has been discussed for the whole region (Mudry and Puig 1991). The NW–SE fault direction is probably a result of the oldest tectonic stress field that has deformed the limestones during the earlier stage of their diagenesis. This is reflected in the reconstructed direction of compression using the stylolite indentation (Shanov and Cousset 1993).

Processes of formation of young NE–SW striking faults are also reflected by the chain arrangement of earthquake epicenters to the northwest of Mont Ventoux (Fig. 2.51). The earthquakes are of low magnitudes (less than $M_L = 3.0$). But, the most important confirmation for the existence of contemporary compressive NE–SW directed strain for this region is the result of the fault-plane solutions from the earthquakes (Shanov and Georgiev 2001). The fault-plane solutions from earthquakes northeast of the studied region show, as a general rule, orientation of the maximum strain compatible to the maximum stress axis of the youngest tectonic stress field, reflected by the joint systems of the limestones (see Table 2.8).

The absolute velocity vectors for the site GRASSE displacement, the nearest Laser Tracking Site, according to the NOVEL1A model and to the Satelite Laser Ranging global geodetic solution, demonstrate practically the same orientation and the velocity range of about 25 mm/year toward northeast and this is the general trend for the western part of the Eurasian Plate (Table 2.8). There is a small residual motion with respect to Eurasia of about 4–5 mm/year in southeastern direction (Shanov and Georgiev 2001). The absolute site motion is in good agreement with the general direction of the *P*-axes from the earthquake fault-plane solutions, and this could be interpreted as an evidence for the direct relationship

Table 2.8 Reconstructed time sequence of the tectonic stress fields for the studied area of Alps Maritimes

Origin	Site №	Post-Eocene (?)	Provencal	Post-Burdiganien	Recent	Method
Earthquake 19.06.1972 M=3.8 N 44.38^0 E 6.35^0					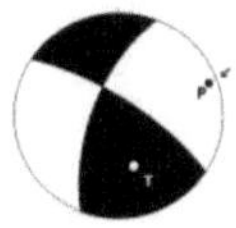	Fault-plane solution
Laser site GRASSE N 43.754464 E 6.921947						Determination of plate motion
1		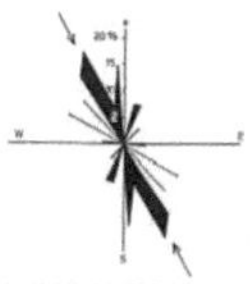				Stylolite's indentation
K_1 – Lower Cretaceous limestones	1	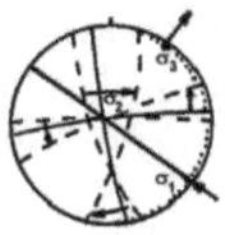	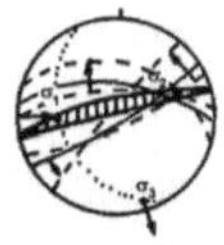			Shear joints
	2		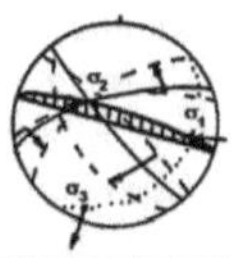	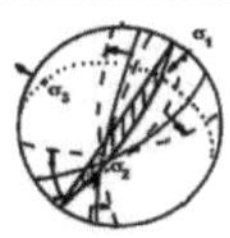		
	3		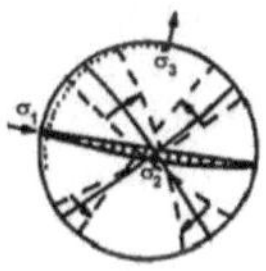	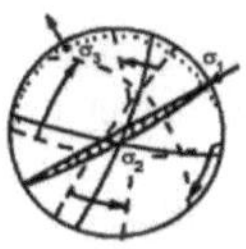		
	4				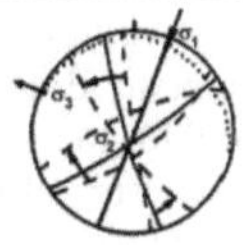	
	5		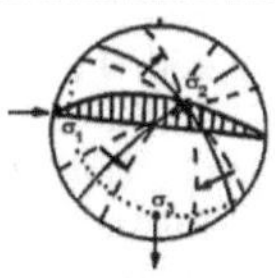			

between the continental motion and the induced compressive strain in the Earth's crust. The rock fracturing under the induced stress, for the case study, opens the way for the underground waters, and the consecutive formation of the karst galleries, the youngest of them following the direction of the maximum tectonic stress axis.

References

Amadei B (1983) Rock anisotropy and the theory of stress measurement. Springer, Berlin, p 478

Andreychuk VN, Benderev AD (1991) Regional peculiarities of the anthropogenic impact on the karst of mountain countries. Informational materials, Ural Branch of Academy of Sciences of USSR, Kongur, p 133 (in Rissian)

Angelier J (1994) Fault slip analysis and paleostress reconstruction. In: Hancock PL (ed) Continental deformation. University of Bristol, Pergamon Press, Oxford, pp 53–100

Angelier J, Mechler P (1977) Sur une méthode graphique de recherche des contraintes principales également utilisable en tectonique et en séismologie: la méthode des dièdres droits. Bull Soc Geol Fr 7:1309–1318

Angelova D, Benderev A, Baltakov G, Ilieva I, Nenov T (1995) On the evolution of the karst in Stara Planina Gorge. J Bulg Geol Soc 3:111–124 (in Bulgarian)

Angelova D, Benderev A, Kostov K (1999) On the age of the caves in the Stara Planina Iskar gorge, NW Bulgaria. In: European Conference "Karst 99", 10–15 Sept 1999, Grands Causses–Vercors, pp 25–35

Antonov H, Danchev D (1980) Ground waters in Bulgaria. Tehnika, Sofia, p 360 (in Bulgarian)

Aubouin J, Brousse R, Lehman JP (1988) Précis de géologie. Tectonique, tectonophysique, 4th edn. Dunod Université, Paris, p 519

Benderev A (1989) Karst and karst waters of Ponor Mountain. PhD Thesis, NIPI-KG, p 157 (in Bulgarian)

Benderev A, Angelova D (1999) Evolution of karst in the southern part of Vitosha Mountain, Bulgaria. Thereotical and applied karstology 11–12(1998–1999):75–82

Benderev A, Shanov S (1997). Karst waters from the region of Bosnek (West Bulgaria): Characteristics and conditions of formation. 12th International Congress of Speleology, La Chaux-de-Fonds. In: Proceedings, v.2, 6 Conference of Limestone Hydrology and Fissured Media, pp 255–258

Benderev A, Shanov S, Ilieva I (1987) Peculiarities of formation of the karst of Western Stara Planina Mountain. In: Proceedings of the conference "The problems of the complex study of the karst of the mountain territories", Tbilisi-Tzhaltubo-Suhumi, 5.12.1987, 5–12.10.1987, pp 94–97 (in Russian)

Benderev A, Shanov S, Ilieva I (2001) Evaluation of the environmental threats affecting the Lakatnik karst region (Western Balkan mountains, Bulgaria). In: Proceedings of the 7th conference on limestone hydrology and fissured media, Besançon, France, 20–22 Sept 2001, pp 31–35

Benderev A, Spassov V, Shanov S, Mihaylova B (2005) Hydrogeological karst features of the Western Balkan (Bulgaria) and the anthropogenous impact. In: Proceedings of the international conference "Water Resources and Environmental Problems in Karst—CVIJIC 2005", Belgrade & Kotor, 13–22 Sept 2005, pp 37–42

Biçoku T, Aliaj Sh (1973a) Historical Tectonic Regionalisation of Albania. In: Proceedings of the Seminar on the Seismotectonic Map of the Balkan Region, UNESCO, Dubrovnik. Appendix: Maps, No.l

Biçoku T, Aliaj Sh (1973b) Neotectonic Map of the Balkan Region, UNESCO. Appendix: Maps, No. 6

Biçoku T, Papa A, Shelu R (1978) Dinarides: Albania. In: Tectonics of Europe and the adjacent territories. Nauka, Moscow, pp 468–472 (in Russian)

Bielenstein HU, Barron K (1971) In situ stress. A summary of presentations and discussions given in Theme I at the Conference of Structural Geology to Rock Mechanics Problems. Department of Energy, Mines and Resources, Mines Branch, Ottawa

Bonchev E (1961) Notes on main faults in Bulgaria. Stud Geol Bulgaria Stratigr tect 2:5–29 (in Bulgarian)

Bonchev E (1971) Problems of Bulgarian geotectonics. Technica, Sofia, p 204 (in Bulgarian)

Brace WF (1960) An extension of the Griffith theory of fracture to rocks. J Geophys Res 65:3477–3480

Cadet JP, Funiciello R (2004) Geodynamic map of the Mediterranean. Commission of the Geological map of the World, 2 sheets

Caputo R (1989) FAULT. A programme for structural analysis. Manual. Department of Earth Sciences, University of Florence, Firenze, p 55

Caputo M, Caputo R (1988) Structural analysis: new analytical approach and applications. Ann Tectonicae II(2):84–89

Caputo M, Caputo R (1989) Estimate of the regional stress field using joint systems. Bull Geol Soc Greece XXIII(1):101–108

Chanov S (1988) Conditions géologiques et tectoniques de la formation des cavités karstiques dans la région de "Barkité" et de "Béliar", montagne de Stara Planina près de Vratza (Bulgarie). Dans: "Récit d'une amitié", Saint Herblain: 12–14

Chanov S, Benderev A (1993) Structural, geophysical and hydrogeological explorations on the karst system of Plateau Guaso (Cuba). In: International symposium on water resources in karst with special emphasis in arid and semi-arid zones, 23–26 Oct, Shiraz, Iran, pp 595–611

Cobiella J (1984) Sobra et origen del extremo oriental de la fosa de Bartlett. Editorial Oriente, Santiago de Cuba, p 43

Cox A, Hart R (1989) Plates Tectonics. Mir, Moscow, p 428 (in Russian)

Depambour Ch, Guendon J-L (2004) Eléments de rèflexion sur la morphogenèse des plateaux de Vaucluse (France): les apports du karst superficiel. Karstologia n° 42

Dilek Y, Koçiu S (2004) Neotectonics, seismology and the mode of Late Cenozoic crustal deformation in the Albanides, Balkan Peninsula. Denver Annual Meeting, 7–10 Nov, paper No 19–14, Abstract

Dreybrodt W (1988) Processes in Karst Systems. Physics, Chemistry and Geology. Springer, Berlin, p 288

Drumia AV, Shebalin NV (1985) Earthquake: where, when, why? Shtiintsa, Kishinev, p 196 (in Russian)

Dunne WM, Hancock PL (1994) Paleostress analysis of small-scale brittle structures. In: Continental deformation. Pergamon Press, Oxford, pp 101–120

Dupin JM, Sassi W, Angelier J (1983) Homogeneous stress hypothesis and actual fault slip: a distinct element analysis. J Struct Geol 15:1033–1043

Engelder T (1994) Brittle crack propagation. In: Continental deformation. Pergamon Press, Oxford, pp 43–52

Evlogiev J, Karachorov P, Todorov M, Hristov D (1997) Study on the genesis of the fractures in Orlova Chuka Cave, Rouse District. Eng Geol Hydrogeol 24(BAS):33–39 (in Bulgarian)

Fage L-H (1981) Spéléo sportive. Edisud, Dans les monts de Vaucluse, p 96

Filipov L (ed) (1992) Geological Map of Bulgaria 1:100 000, Sheet "Byala"

Gamond JF (1983) Displacement features associated with fault zones: a comparison between observed examples and experimental models. J Struct Geol 5:33–45

Georgiev I, Kotzev V (1995) Tectonic plate motion from satellite laser ranging to Lageos 1. In: Proceedings of the I-st international symposium on deformations, 5–9 Sept, Istanbul, Turkey, pp 454–472

Georgiev Tz, Shanov S (1991) Contemporary geodynamics of the western part of the Moesian Platform (Lom Depression). Bulg Geoph J LXVII(3):3–9 (in Bulgarian)

Georgiev I, Dimitrov D, Belijashki T, Pashova L, Shanov S, Nikolov G (2007) Geodetic constraints on kinematics of southwestern Bulgaria from GPS and leveling data. Geological Society, London, Special Publications 01/2007; 291(1):143–157. doi:10.1144/SP291.7

Gephard J, Forsyth D (1984) An improved method for determining the regional stress tensor using earthquake focal mechanism data. J Geophys Res 89:9305–9320

Gochev P (1983) Subhercynian Allochthon in West Srednogorie. Geotect Tectonophys Geodyn Geol Inst BAS 15:3–17 (in Bulgarian)

Gouvernet C, Guieu G, Rousset C (1971) Provence. Guides géologiques régionaux, Masson et Cie, Paris, p 230

Griffith AA (1924) Theory of rupture. In: Proceedings of first international congress for applied mechanics, Delft, pp 55–63

Harta Geologjike RPS Te Shquiperise M 1:200 000 (Geological Map of Albania) (1981) Tirana, Ministria e Industrise dhe e Minerave

Haydutov I, Dimitrova R (ed) (1992) Geological map of Bulgaria 1:100 000, Sheet "Berkovitsa"

Hmelevskoy VK (1984) Electrical prospecting. University of Moscow, Moscow, p 422 (in Russian)

Hunt L, Reynolds S, Hadley S, Downton J, Chopra S (2011) Causal fracture prediction: curvature, stress and geomechanics. Leading Edge SEG 30(11):1274–1286

Hydrogeological Map of PSR of Albania M 1:200 000 (1981) Tirana

Ilieva I, Zdravkov I, Benderev A (1981) The role of the tectonic movements for the development of the karst at the area or Cherepish Railway Station. Proceedings "Scientific and Practical Conference on Tourism, Alpinism, Caving and Environment Protection", Rousse, 4–6 May 1979, pp 207–214 (in Bulgarian)

Jaeger J (1975) Rock mechanics and engineering. Mir, Moscow, p 256 (in Russian)

Jelev V (1982) Characteristics and development of Vitosha central magmathogenetic structure. PhD Thesis, University of Mining and Geology, Sofia (in Bulgarian)

Karagjuleva J (1971) The northern boundary of Fore-Balkan. In: Tectonics of Fore-Balkan. BAS, Sofia, pp 66–107 (in Bulgarian)

Kastelic V, Radulov A, Glavanovic B (2011) Improving the resolution of seismic hazard estimates for critical facilities: the database of Eastern Europe crustal seismogenic sources in the frame of the SHARE project. In: Proceedings of the 30th meeting of the GNGDS (Gruppo Nazionale di geofisica della terra solida), pp 218–221

Kostadinov V (1965) The border between Kaishte and Srednogorie tectonic zones and stages of development. In: Proceedings CBGA, VII Congress, vol 1, pp 45–49 (in Russian)

Kostov K (1997) Karst morphology in Bazovski part of Vratsa mountain (Stara planina, NW Bulgaria). In: Proceedings of 12th International Congress of Speleology, La Chaux-de-Fonds (Newchatel, Switzerland), 10–17 Aug, vol 1, pp 409–412

Letouzey YJ (1986) Cenozoic paleo-stress pattern in the Alpine foreland and structural interpretation in a platform basin. Tectonophysics 132:215–231

Lilienberg DA (1983) Morphotectonics and recent geodynamics of the interaction region between Cuba microplate and Bartlett deepwater trench. In: Symposium of the morphotectonics Working Group IOU, Bulgaria, pp 204–232

Lundborg H (1968) Strength of rock-like materials. Int J Rock Mech Min Sci 5(5):427–454

Markowicz M, Popov V, Pulina M (1972) Coments of Karst denudation in Bulgaria.—Geogr. Polonica 23:118–139

Matova M, Angelova D (1994) Geological, seismotectonical and environmental problems of Pernik graben (Western Bulgaria). In: Proceedings of 7th Congress of the International Association of Engineering geology, Lisbon, Portugal, vol III, pp 2027–2032

Mattauer M, Mercier JL (1980) Microtectonique et grande tectonique. Livre Jubilaire de la Soc Géol Fr Mem h.-s. n° 10:141–161

McCarthy D (ed) (1996) IERS Technical note 21: IERS Convention, 95 p. Observatoire de Paris

McClintock FA, Walsh JB (1962) Friction on Griffith cracks in rocks under pressure. In: Proceedings of IV U.S. National congress of applied mechanics, Berkeley

Meyer B, Sébrier M, Dimitrov D (2007) Rare destructive earthquakes in Europe: The 1904 Bulgaria event case. Earth Planet Sci Lett 253:485–496

Michael AJ (1987) Use of focal mechanisms to determine stress—a control study. J Geoph Res 92:357–368

Mihailova B, Mitev A, Benderev A, Shanov S (2006) Hydrogeological and geophysical investigations for localization the river recharge of the ground waters of the Bosnek Karst Region. In: Proceedings National conference "Geosciences", 30 Nov–1 Dec 2006, Sofia, pp 403–406 (in Bulgarian)

Mishev D, Popov V (1958) The karst of Vratsa Mountain, Sofia. Nature 7:7–13 (in Bulgarian)

Mudry J, Puig JM (1991) Le karst de la Fontaine de Vaucluse. Karstologia 18:29–38

Muço B (1994) Focal mechanism solutions for Albanian earthquakes for the years 1964–1988. Tectonophysics 231:311–323
Nikolaev PN (1977) Methode of statistical analysis of joints and reconstruction of the tectonic stress fields. J High Schools Geol Prosp Moscow 12:103–115 (in Russian)
Nystuen (ed) JP (1989) Rules and recommendations for naming geological units in Norway by the Norwegian Commitee on Stratigraphy. Nor Geol Tidsskr 69(Supplement 2):111
Paskalev M, Benderev A, Shanov S (1992) Tectonic conditions of the region of the Iskrets Karst Springs (West Stara-Planina). Rev Bulg Geol Soc 53(part 2):69–81 (in Bulgarian)
Popov V (1964) Morphology and genesis of Ledenika Cave. Bull Geogr Inst BAS 8:77–87 (in Bulgarian)
Popov V (1982) Travel under the ground. Nauka i Izkustvo, Sofia, pp 152 (in Bulgarian)
Price NJ (1959) Mechanics of jointing in rocks. Geol Mag 96:149–167
Price NJ (1966) Fault and joint development in brittle and semi-brittle rock. Pergamon Press, Oxford
Radev J (1915) Karst forms in Western Stara Planina. Ann Sofia University, Hist Phil Fac 10–11:149 (in Bulgarian)
Radulov A (2002) Quaternary karst in the valley of Rusenski Lom River and its tributaries. PhD Thesis, Geological Institute, BAS (in Bulgarian)
Radulov A, Vanneste K, Verbeeck K, Yaneva M, Petermans T, Camelbeeck T, Shanov S (2006) Paleoearthquake correlation in three trenches along Orizovo—Chirpan active fault segment. National conference "Geosciences 2006", Sofia, 30 Nov–1 Dec, Proceedings, pp 71–74
Radulov A, Dilov Tz, Yaneva M, Nikolov N, Kostov K, Nikolov V (2011) Earthquake faults in Sofia Basin. In: National conference "Geosciences 2011", Sofia, Proceedings, pp 99–100 (in Bulgarian)
Radulov A,Yaneva M, Shanov S, Nikolov V, Kostov K, Nikolov N, Hristov V, Dobrev N, Mitev A (2012) Report on field survey after May 22, 2012 Pernik earthquake, June 6, 2012, Part 1: Coseismic geological effects related to May 22, 2012 Pernik earthquake, Western Bulgaria http://www.geology.bas.bg
Ramsay JG (1980) Shear zone geometry: a review. J Struct Geol 2:83–99
Reid HF (1911) The elastic rebound theory of earthquakes. Bull Dept Geol Univ California 6:413
Rojas-Agramonte Y, Neubauer F, Handler R, Garcia-Delgado DE, Friedl G, Delgado-Damas R (2003) Neogene-Quaternary tectonics along the North Caribbean transform fault, Cuba. EGS—AGU—EUG Joint Assembly, Abstracts from the meeting held in Nice, France, 6–11 Apr 2003, abstract #2830
Sassi W, Carey-Gailhardis E (1987) Interpretation mechanique du glissement sur les failles: introduction d'un critere de frottement. Ann Tectonicae I(2):139–154
Scanvic J-Y, Weecksteen G (1984) Carte des linéaments de la France. Echelle 1:1 000 000, BRGM
Scholz K (2002) The mechanics of earthquake faulting. Cambridge University Press, Cambridge, p 472
Scorpil H, Scorpil K (1895) O kraskych zjevech v Bulharsku. Rozpr Ceske Akad 4(2):1–35
Scorpil H, Scorpil K (1898) Sources et pertes des eaux en Bulgarie. Mem Soc speleol 15:99–139
Shanov S (1999) Tectonic conditions for the karst processes near Karlukovo Village (Fore-Balkan). In: Proceedings National scientific conference on the problems of Karst and Speleology, Sofia, pp 37–43 (in Bulgarian)
Shanov S (2005) Post-Cretaceous to recent stress fields in the SE Moesian Platform (Bulgaria), Tectonophysics, vol 410, Issues 1–4, The Carpathians-Pannonian Basin System—Natural Laboratory for Coupled Lithospheric-Surface Processes, 9 Dec 2005, pp 217–233
Shanov S, Benderev A (2005) Seismic impact from earthquakes and from grouped blasts in quarries on the discharge of karst springs. In: Proceedings of the international conference "Water Resources and Environmental Problems in Karst—CVIJIC 2005", Belgrade & Kotor, 13–22 Sept 2005, pp 145–150
Shanov S, Cousset O (1993) Tectonic stress fields and Karst processes in the Plateau of Vaucluse (South-Eastern France). In: International symposium on water resources in Karst with special emphasis in arid and semi arid zones, 25–26 Oct, Shiraz, Iran, Proceedings, vol 2, pp 775–783
Shanov S, Georgiev I (2001) Tectonic stresses and their control on the karst formation in the Plateau of Vaucluse (South-East France). Riviera 2000—Tectonique active et géomorphologie. Villefranche-sur-Mer, Revue d'Analyse Spatiale, No spécial 2001:139–144
Shanov S, Georgiev T, Dimitrov B (1988) Contemporary tectonic stress field in North Eastern Bulgaria. J Bul Geol Soc 1:39–46 (in Bulgarian)
Shanov S, Stoyanov S (1986) On a method for treatment and dynamic interpretation of joints. Rew Bulg Geol Soc 57(part 1):64–73 (in Bulgarian with an Abstract in English)
Shanov S, Benderev A, Ilieva (2001) Traces of recent active tectonics in the cave of Duhlata (Southwestern Bulgaria). Riviera 2000—Tectonique active et géomorphologie. Villefranche-sur-Mer, Revue d'Analyse Spatiale, No spécial 2001:145–150
Sharma PV (1997) Environmental and engineering geophysics. Cambridge University Press, Cambridge, p 475
Sheriff RE (1984) Encyclopedic dictionary of exploration geophysics. Nedra, Moscow, p 352 (in Russian)
Spasov V, Benderev A, Gabeva D (1998) Perspectives for more effective using of the underground karst waters of the region of the town of Vratsa. In: Proceedings of the National conference "Water resources—usage and preservation", Sofia, 23–25 Sept 1998, pp 47–51 (in Bulgarian)
Spenser EU (1981) Introduction in Structural Geology. Nedra, Leningrad, p 368 (in Russian)
Stacey F (1972) Physics of the Earth. Mir, Moscow, p 342 (in Russian)
Stoyanov S (1970) Certain physical aspects of faulting in the Earth's crust. Bull Geol Inst Ser Geotectonics BAS 19:127–140 (in Russian with abstract in English)
Trollop DH (1983) Criteria of failure of brittle rocks. In: Introduction in rock mechanics. Mir, Moscow, pp 96–113 (in Russian)
Tronkov D (1965) Tectonics and analysis of the structures in the Vratza Block of Western Stara Planina. Plastic deformations near the fault planes. Stud Geol Bulgaria Ser "Stratigraphy and Tectonics" 6:217–257 (in Bulgarian)
Tsankov Ts (ed) (1991) Geological map of Bulgaria 1:100 000, Sheet "Vratsa"

Tzankov T, Burchfiel C, Royden L (1998) Neotectonic (Quaternary) Map of Bulgaria. Scale 1:500 000. General Aspects on the Quaternary Tectonic Evolution of Bulgaria. Publishing House Grafika, Sofia
Uhov SB (1975) Rock basement of hydro-technical facilities. Energia, Moscow, p 264 (in Russian)
USGS (2004) Shuttle Radar Topography Mission, 3 Arc Second scenes N43E023 and N43E024, Unfilled Unfinished 2.0. Maryland, University of Maryland, College Park, Global Land Cover Facility, Feb 2000
Vanneste K, Radulov A, De Martini P, Nikolov G, Petermans T, Verbeeck K, Camelbeeck T, Pantosti D, Dimitrov D, Shanov S (2006) Paleoseismologic investigation of the fault rupture of the 14 April 1928 Chirpan earthquake (M 6.8), Southern Bulgaria. J Geophys Res 111:B01303. doi:10.1029/2005JB003814
Vapcarov I, Galabov J, Michev K, Georgiev M, Vrablianski I (1974) Neotectonic map of Bulgaria, Scale 1:000 000. In: Proceedings of the seminar on the seismotectonic map of the Balkan region, UNESCO, Skopje, App. Maps, No 11
Vergely P, Sassi W, Carey-Gailhardis E (1987) Analyse graphique des failles a l'aide des focalisations des stries. Bull Soc Geol de France 8, t. III, n° 2:395–402
Vogt J, Godefroy P (1981) Carte sismotectonique de la France. Présentation et mode d'emploi. Mem. du BRGM, No 111
Vrablianski B (1974) Main lines of tectonic activation of the Earth's crust in Bulgaria during the Antropogean. Compte Rend. de l'Acad Bulg des Sci 27(7):953–956
Yakubovskiy Y (1973) Electrical prospecting. Nedra, Moscow, p 299 (in Russian)
Zagorchev I, Marinova R, Tchounev D, Tchumachenko P, Sapunov Y, Yanev S (1994) Explonatory note to the Geological map of Bulgaria in scale 1:100000, Pernik map sheet Committee of Geoloy and Mineral Resources, Sofia, p 92 (in Bulgarian)

3 Recent Geodynamics and Karst

3.1 Traces of Paleoseismicity and Active Tectonics in Karst: Historical Notes

The first to discuss damages of speleothems in caves as a result of tectonic events (earthquakes) was Hohenwart (1832) in his observations of Postojna Cave in the first half of nineteenth century. According to the translation of Kempe (2004), Graf von Hohenwart wrote in his guide of the cave: "Also the forceful changes in the grotto show that nature rules after own law -unknown to us—in this netherworld.

The common opinion about these alterations in such underground caves is that weak vaults cavein due to earthquakes and that loose stems fall from above and that those stalactite columns which do not adhere strongly enough at their junction to the ceiling are precipitated. The careful observer, however, discovers that many alterations cannot be explained by this common opinion." (Kempe 2004, p. 272).

The hypothesis that speleothems can "record" seismotectonic history is mentioned in the studies of Becker on the caves Bing Hohle in Germany and Han-sur-Lesse, Belgium (Becker 1929). Damages of speleothems are also reported by Spöcker in Franken Jura but without assumption for their seismic nature (Spöcker 1933).

Later, Schillat established an availability of broken speleothems in Langenfeld Cave and discussed the seismic origin of the deformations (Schillat 1965, 1970). Systematic studies on deformed stalagmites in Postojna Cave are published by Gospodarich (1968, 1977, 1981). Speleological phenomena with possible tectonic origin in karst areas of the Dinaric Mountains are investigated by Garasic (1981) and Garasic and Cvijanovic (1985).

During the 1970s, similar observations in the karst regions of Crimea were made by Dublyansky and Molodih (1972). Professor Dublyansky distinguishes five groups of karst features bearing seismotectonic information: displaced cave galleries, moved limestone blocks, corrosion-gravitational forms (caves and pot-holes of tectonic origin), cave breakdowns, and speleothems (Dublyansky 1995).

Moser and Geyer accepted the presence of deformed speleothems in caves in the Bavarian Upper Franconia region, Germany, as a result of strong earthquakes (Moser and Geyer 1979).

The paleoseismologic research in karst became more popular in Europe after a number of studies of caves in Italy were carried out in the 1980s by the team of Prof. Paolo Forti at the University of Bologna (Forti 1997, 1998; Forti and Postpischl 1981a, b, 1984, 1985, 1986, 1987; Forti et al. 1983; Postpischl et al. 1990, 1991).

Undoubtedly, it is almost impossible within the frames of this book to examine in detail all studies conducted till date. In this chapter, we try to provide a brief overview of the paleoseismologic and active tectonics research carried out so far in caves from different karst regions of the world.

Belgium is characterized by relatively low contemporary seismic activity. The strongest historical earthquake was in September 18, 1692 in Verviers, Northern Ardennes (Camelbeeck 1998). However, several trench studies along the active Bree fault in northern Belgium (Vanneste et al. 2001) were performed obtaining further evidence for three strong

S. Shanov and K. Kostov, *Dynamic Tectonics and Karst*, Cave and Karst Systems of the World,
DOI: 10.1007/978-3-662-43992-0_3,

Fig. 3.1 Konstantin Kostov pointing out broken stalagmites recovered with new speleothems in Hotton Cave, Belgium (*photo* Serge Delaby)

earthquakes in the last 40,000 years, the latter being dated between 610 and 890 BC.

The paleoseismologic research in the Belgian caves was made by the team of the Centre for Study and Applied Research on Karst (CERAK) at the Polytechnic Faculty of Mons. The investigations undertaken at the initiative of Prof. Yves Quinif are important in methodical aspect as introducing terms such as "seismothems" and "stalagmite cemeteries" (Quinif 1996, 1999, 2000).

In rarely visited parts (Reseau Sud) of the longest cave system in Belgium, Han-sur-Lesse, have been identified a large number of fallen stalagmites. The U-series dating shows the influence of two strong earthquakes in the Holocene and Late Pleistocene (Quinif 1999).

Serge Delaby continued the research in the cave Hotton. At the third highest level of the fossil cave discovered in 1958 are conducted precise mapping and measuring of the size and spatial orientation of the broken stalagmites (Delaby 1999, 2000) (Fig. 3.1).

Important studies of breakdowns in caves (archeological sites) in *France* were held by Blanc. Multiphase activation of breakdowns caused by seismic situation is established applying a complex analysis. Blanc performed a seismic microzoning to determine safe places for building nuclear power plants in Southern France (Blanc 1985).

L. Bruxelles, J. L. Guendon, and Y. Quinif found an array of fallen stalagmites, broken stalactites, and fractured flowstones dislocated by active faults in the pot-hole Portalerie (−149 m) in Southern France. The morphological analysis of deformations suggests the impact of at least two strong paleoearthquakes. The uranium-thorium dating shows that seismic events occurred in the time interval 36,800–4,500 BP (Bruxelles et al. 1998).

The U-series dating of broken and calcified thin stalactites (soda straws) in the cave Ribiere performed by the Franco-English team demonstrates the impact of a Pleistocene seismic event aged 170,000–190,000 BP (Delange and Guendon 1998).

The team of Pons-Branchu conducted research in Salamandre Cave in southeastern France, characterized by many fallen and broken speleothems covered with new generation of calcite deposits. The carried TIMS U-Th dating of two fallen stalagmites recovered with new formations suppose the influence of paleoseismic events aged 7 ± 0.35 ka and between 1.1 ± 0.1 and 6.3 ± 2 ka (Pons-Branchu et al. 2004).

Very important studies were undertaken by Eric Gilli and his colleagues from the Center for the Study of Karst in Nice (Gilli 1986, 1995, 1996; Gilli et al. 1998) in caves in France and abroad. In the Observatoire Cave in Exotic Garden of Monaco is committed a hand drilling. In the shallow well with depth of 23 cm are identified three generations of broken thin stalactites as the first (youngest) one meets the Ligurian earthquake of 1887 ($M_s = 6$–6.5) and the third (oldest) lies on buried calcite crust aged 35,000 BP (Gilli 1998).

The group of Gilli carried out successful speleopaleoseismologic observations in the Asian part of *Turkey*. Studies on the Tilkiler Cave near the Oymapinar Dam (Manavgat, Taurus Mountains, southern Turkey) found plenty of broken stalactites on the floor of the cavern. Some of them were driven vertically into the sandy sediments, and elsewhere described earlier clusters of broken stalactites, together with calcite crust. The cave was discovered shortly before the visit of Gilli, during the digging of a tunnel and had no natural entrance. The performed radiocarbon dating indicates that fractures-aged stalactites in Tilkiler are of age 5,000–800 BP (Gilli 1995).

Gilli performed research in karst areas of *Costa Rica* and *USA* as well. Costa Rican territory is characterized by very high seismicity—since 1940 registered ten earthquakes with magnitude of more than 7 (Gilli 1995). Extensive paleoseismologic studies of the caves were undertaken in 1994 by the team of Gilli. There have been many different forms of deformation caused by active faults (Coredores Cave to Peninsula Osa) and centimetric displacements of the cross section of cave galleries. Inclined stalagmites are described in the cave Tersiopelo—probably a result of the earthquake in 1991 ($M_s = 7.1$) (Gilli 1995).

In the *United States* deformed speleothems are found in the caves, Sutherland Peak (Southern Arizona) Endless Cave, Sand Cave, MacKittrick Cave, Carlsbad Caverns, and Hidden Cave (New Mexico). In Sutherland Peak is described broken stalactites—probably resulting from the impact of Sonora earthquake (Mexico, May 3, 1887, $M = 7.2$). Gilli visited caves in Carlsbad District of New Mexico characterized by the presence of broken speleothems and breakdowns. The research was hampered by strong anthropogenic interference, dating from the second half of the nineteenth century in the Carlsbad Caverns National Park (Gilli and Serface 1999).

Panno from the Illinois State Geological Survey and his coauthors published a study on caves located within 250 km of the New Madrid Seismic Zone where established stalagmites with deviated axes of grown and fallen stalagmites recovered with active new stalagmites. The data from the accomplished U/Th mass spectrometric dating of three speleothems show the influence of single seismic event (Panno et al. 2009).

Research in the Brujas Cave in the southern Andes, *Argentina* was carried by Sancho et al. (2003). In hardly accessible passages are identified clusters of broken stalactites, which are interpreted as a sign of a seismic event with Pre-Holocene age. Franko Urbani reported for rotated rock blocks and fractured speleothems as indices for paleoseismicity in Guanasna Caves near Caracas, *Venezuela* (Urbani 2002).

The method of Prof. Paolo Forti from the University of Bologna, *Italy*, was founded on the belief that stalagmites are natural pendulums and respond to changes in the cave floor. Selected stalagmite samples are cut longitudinally and the variance of their growth axis is interpreted as evidence of the impact of earthquakes. The result of the comparison between the time series variations in stalagmite sections and corresponding catalogs of earthquakes represents a clear correlation. For example, in stalagmites from the caves Buko dei Buoi and Spinola in the vicinity of Bologna is established abrupt deviation from their axes of growth, corresponding to the earthquake of January 3, 1117—the strongest in northern Italy. In the studied stalagmites are found traces of three seismic events earlier than 1117 (Forti and Postpischl 1984, 1987).

Within the last years, tectonic studies in karst in *Slovenia* were generally performed by Dr. Stanka Šebela and her collaborators from the Karst Research Institute in Postojna. In 2004, four extensometers were installed in caves and two of them mounted in Postojna Cave (Gosar et al. 2011; Šebela 2008, 2009, 2010; Šebela et al. 2005).

Pérez-López and others (2009) found a coseismic ceiling block collapse occurred at −156 m in Benis Cave (−213 m, SE *Spain*), associated with the Mula earthquake (1999, $M_b = 4.8$, VII MSK). Their analysis suggests that an active fault segment determined the morphology of the cave, where strong paleoearthquakes ($6 \leq M \leq 7$) took place. As a consequence of this intense seismic activity, a cave gallery collapsed and a new seismothem was recognized (Pérez-López et al. 2009).

The intensity of one of the most destructive earthquakes in *Switzerland* and Western Europe (Basel earthquake, October 18, 1356) is measured between VIII and IX MSK. French and Swiss scientists take the search for traces of the earthquake in the karst area around Basel (Lemeille et al. 1999). Paleoseismologic traces are encountered in the caves Dieboldslöhli and Bäterloch, situated 10–15 km from the epicenter of the earthquake of 1356. In the caves are described stalagmites with anomalous growth, broken stalactites, rock falls. The results of dating of six stalagmite samples by U/Th alpha spectrometry show the young age of the deformation—age up to several millennia (Lemeille et al. 1999).

During the last decade within the structure of the Swiss Seismological Survey to ETH—Zurich was created a research team dealing with paleoseismologic studies of caves. Lacave and colleagues explored the Milandre Cave, North Jura and observed a major deformation of stalactites and stalagmites, after which analysis the impact of a strong earthquake in the Upper Pleistocene was supposed (Lacave et al. 2004).

The paleoseismic studies in karst of *Israel* are related mainly to the work of Dr. Elisa J. Kagan from the Institute of Earth Sciences at the University of Jerusalem and her colleagues from the Geological Survey of Israel (Kagan 2002a, 2002b; Kagan et al. 2002a, 2002b, 2005, 2006). Kagan studied the caves Soreq and Har-Tuv near Jerusalem, located 60 km west of the Dead Sea Transform fault. A variety of deformations with seismotectonic origin are established: fallen stalagmites with a new generation of calcite deposits on them (38 cases) and breakdowns covering old speleothems and in turn covered with new stalagmites.

The performed 230Th/234U dating by mass spectrometry defined four distinct phases of intense seismotectonic activity during the last 20,000 years (Kagan et al. 2002a, 2005).

The studies of Yael Braun on speleothem deformations in Denya Cave near Haifa and their dating show the influence of two seismic events: 4.8 ± 0.8 ka and 10.4 ± 0.7 ka. The results from Denya Cave located close to Carmel fault are good correlated with the obtained paleoseismic data from the caves near the Dead Sea Transform fault (Braun et al. 2010).

Coseismic damages in Gunung Sitoli Cave in *Indonesia* connected to the March 5, 2005 Nias earthquake ($M = 8.7$) are reported by Omer Aydan. Various forms of damage were observed: stalactites failings, stalactone ruptures, roof collapses, etc. (Aydan 2008).

3.2 Methods of Study

To our knowledge, till now no unified complex methodological apparatus for dynamic tectonics studies in karst is approved. This book is an attempt to put into practice such methodology using comprehensive approach with application of techniques from a number of modern methods of paleoseismology, engineering seismology, karstology, absolute geochronology, geophysics, geology, and geomorphology. The principle of complexity is a precondition for introducing into the practice of coherent set of specialized methods for dynamic tectonics research in karst. A specific feature in the methodical aspect is the substantial fieldwork under difficult conditions, which must be carried out using caving techniques.

3.2.1 Morphological and Statistical Analysis of Deformed Speleothems

The method of statistical analysis of the spatial orientation of deformed, fallen, and broken speleothems is based on the assumption that the availability of clearly expressed preferred directions of the studied speleothems is a precondition to the influence of coseismic effect on cave sediments.

Probably the first attempt for statistical analysis of deformed speleothems as indices of strong seismic events in karst areas was performed in the former USSR by Prof. Dublyansky at the beginning of the 1970s. His studies on the preferred directions of fallen stalactones in Crimean caves and pot-holes established a relation to the epicenters of the strongest historical earthquakes in the Crimean peninsula. The rose-diagram of the orientation of 62 broken columns in 14 caves shows that they are connected to Feodosia-Sudak (40 %), Yalta (38 %), Sevastopol (12 %), and Alushta epicenter zones (10 %) (Fig. 3.2) (Dublyansky 1995).

Gospodarich established a correlation between the collapsed speleothems in Postojna Cave and the seismicity in Slovenia (Gospodarich 1977, 1981).

During the last two decades, this methodology is developed in several papers of Dublyansky (1995), Gilli (1995, 2005), Gilli et al. (1998), Quinif (1997), Delaby (1999, 2000), Kagan (2002a, b), Kagan et al. (2005), Kostov (2000, 2001, 2002, 2004, 2008), etc. As an

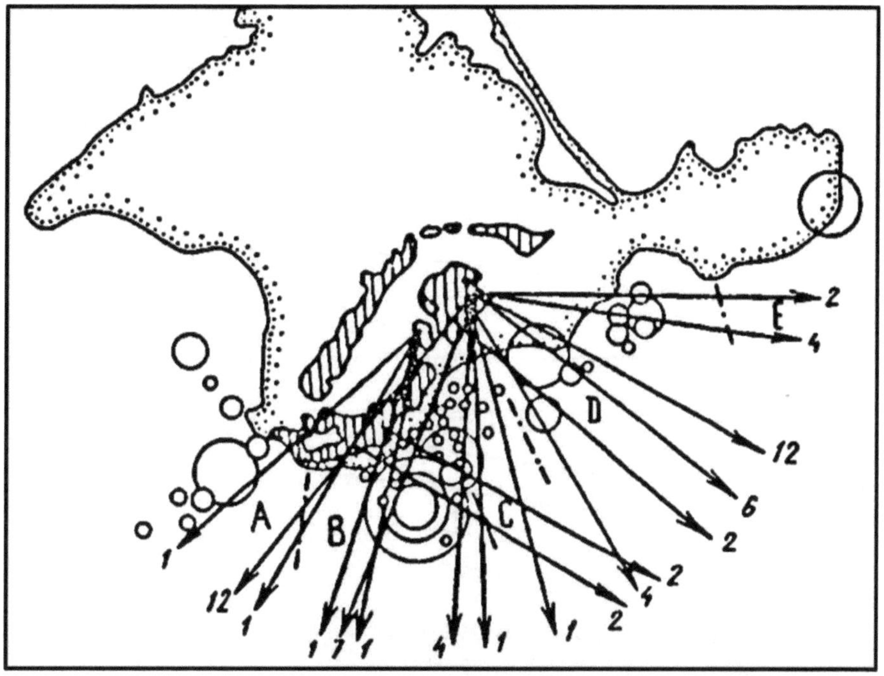

Fig. 3.2 Directions of the fallen stalactones in caves in the Crimean Peninsula. The number of stalactones in the relevant direction are presented numerically. Epicentral zones: A—Sevastopol, B—Yalta, C—Alushta, D—Feodosia-Sudak, E—Kerch-Anapa (Dublyansky 1995)

indicator of paleoseismic events Dublyansky uses fallen stalactones, Gilli—broken stalactites, and Quinif and Delaby—fallen or inclined stalagmites. In our opinion, a complex approach with measuring the directions of the entire available range of deformed speleothems increases the reliability of the obtained results.

In some cases, the breaking or inclination of the speleothems may be connected to other reasons: subsidence of clay deposits beneath speleothems, exaration, biogenic, or anthropogenic influence.

(A) *Biogenic or anthropogenic impact on the speleothems;*

With the appearance and development of caving and cave tourism arise the problem of the protection of cave landscapes. The first show cave considered in Europe was Vilenica in Slovenia, where an entrance fee was introduced for visitors in 1633. The traces of negative human impact in the caves are presented most often by broken stalactite and stalagmite fragments and are typical for the easily accessible parts. Their determination is not difficult in field studies.

Becker et al. (2006) reported that in the Dograrati Cave (Kefalonia, Greece) the stalactites were broken by gunfire during WW2.

The rare cases of negative biogenic impact on cave sediments are represented by breaking of stalagmites by large Pleistocene mammals. Gilli reported

Fig. 3.3 Serge Delaby near massive “hanging” stalagmite without fluvial deposits under his base in Troana Cave, Bulgaria. The speleothem is close to cave river

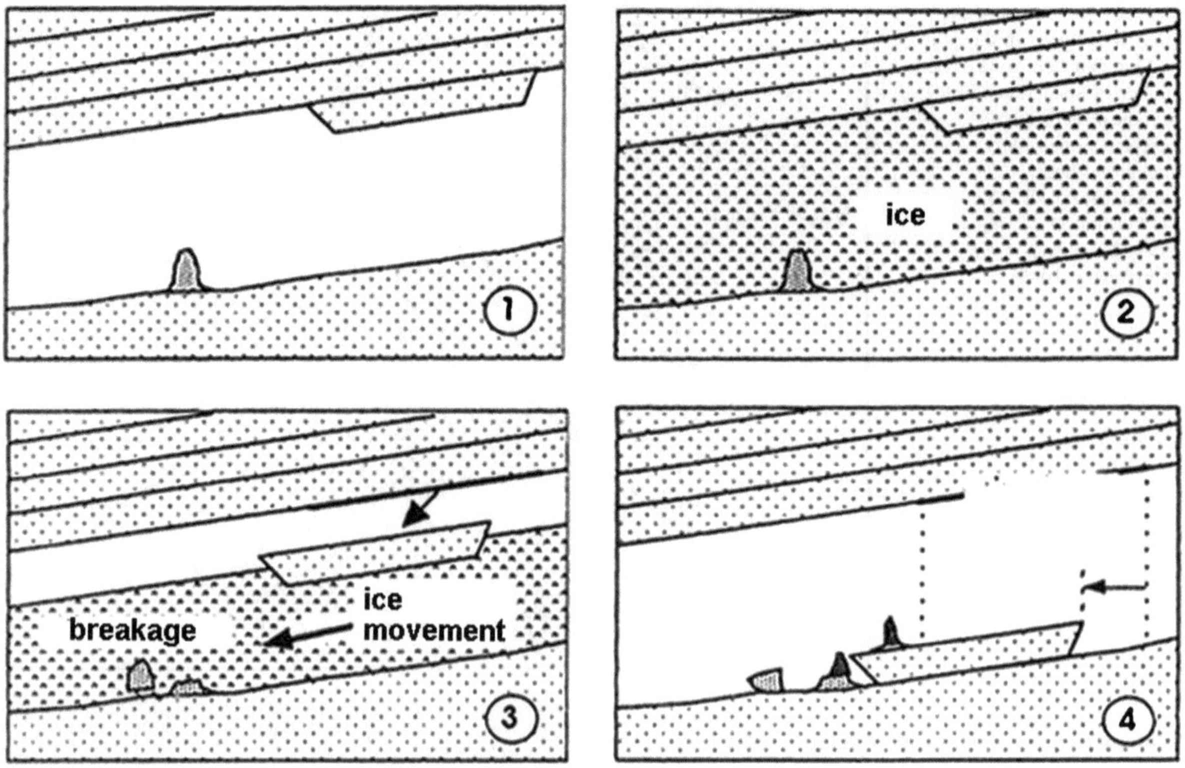

Fig. 3.4 Deformations of speleothems by exaration activity: *1* Initial position; *2* Filling the gallery with ice; *3* Movement of the glacier with breaking of a stalagmite; *4* Present situation (Gilli 1999)

strangely deformed massive speleothems in Pyrenean caves—"cemeteries" of a cave bear. The deformations (fallen stalagmites) near bear dens can be interpreted as a result of the activities of *Ursus Spelaeus*. (Gilli 1995).

(B) *Effects of subsidence or landslides of fluvial cave sediments*;

The fluvial cave deposits are represented by sands, shales, and siltstones. Depending on the morphology of the cave and the conditions of sedimentation, their thickness can possibly exceed 7–8 m. In caves with active hydrodynamic regime, slow movement of these sediments, subsidence, and landslides in some passages near the cave river can be observed (Fig. 3.3). In case of formation of speleothems on fluvial deposits, slow inclination or failing is possible.

In the French cave is established very rare breakage of stalactites by mud-stone torrent (Vanara 1997). The preconditions for the emergence of such a catastrophic geomorphologic phenomenon are the following: (1) The cave is a shallow hole (ponor) with a narrow entrance and a minor section of galleries; (2) The cave is situated at the lowest point of a large karst depression (polje, blind valley, or uvala) where it is possible to accumulate a barrage of devastating impacts.

(C) *Impact of ice movement*;

Dr. E. Gilli from the Center for Karst Studies in Nice presented the hypothesis that in some caves, speleothems breaking is possible as a result of the movement of underground Würm glacier (Gilli 1999) (Fig. 3.4). The model study of Lundberg and McFarlane (2012) for Kents Cave, Devon, UK, shows that up to 3 m thick sediments can be frozen in constant low temperatures of −10 to −15°. The freezing could fracture flowstones up to 13 cm thick. Nowadays, the idea of exaration effect on the cave's morphology and sediments is widely accepted by many karstologists (Kempe 2004; Becker et al. 2006). In our opinion, this hypothesis is not applicable for all cave areas and is suitable for karst massifs located near the snow line of the last glacial period.

The precise analysis and logical ignoring of the described possible causes of deformation of

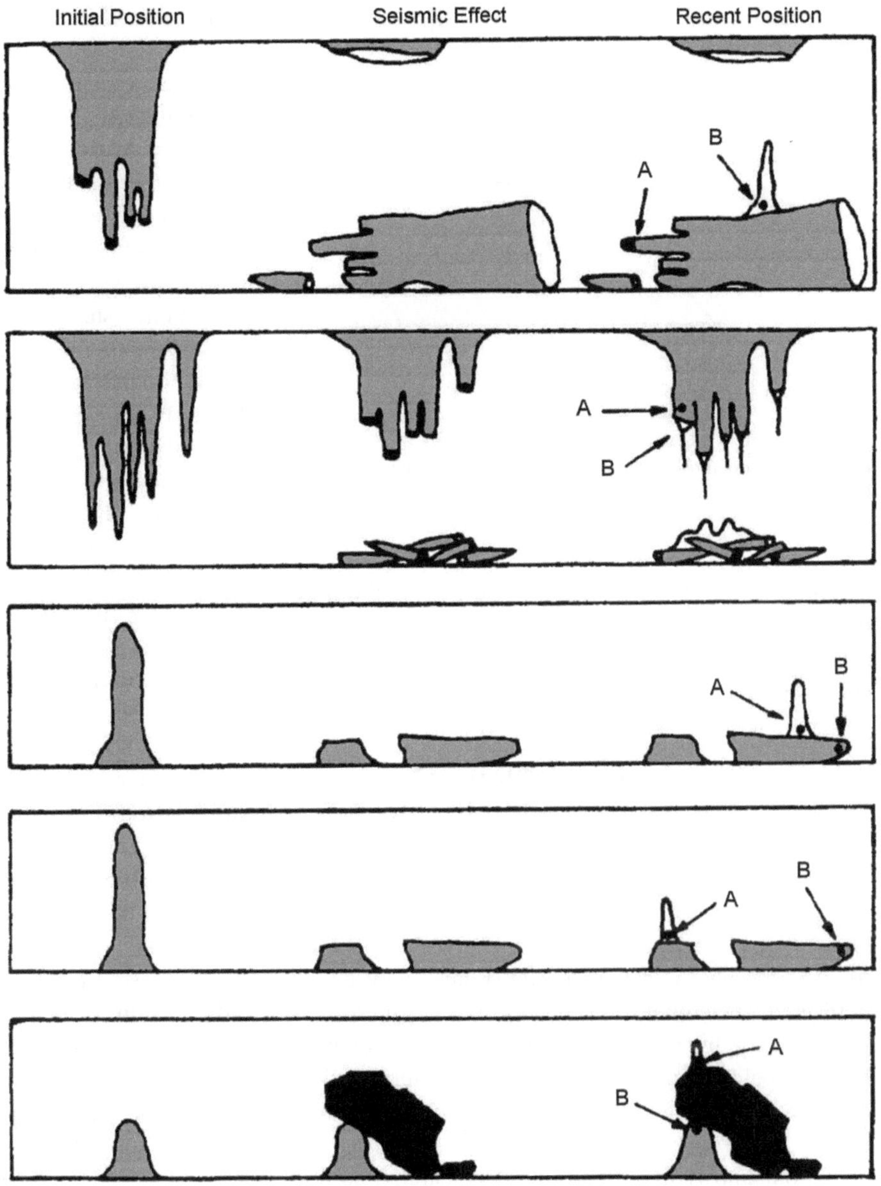

Fig. 3.5 Morphological types of seismode formations of speleothems. *A* and *B*: Places for dating samples for the purpose of establishing the age of the seismic event (modified after Dubois and Grellet 1997)

speleothems lead to the last possible reason—seismotectonic effect. Therefore, to carry out correct statistical analysis of deformed speleothems it is necessary to have the following elements:

1. The studied cave should be well protected from negative anthropogenic interference—i.e., relatively inaccessible and unvisited;

2. In the studied cave is found a sufficient number of deformed speleothems, allowing statistical analysis;

3. Established deformed secondary cave formations are in dry areas of the fossil cave. In the presence of an underground river, to analyze samples in remote parts of it;

4. Cave is a narrow cavern (sinkhole) draining a considerable territory;

5. Deformed or fallen bodies lie on a stable horizontal surface, which excludes their secondary redeposition;

6. The analyzed deformed specimens are calcified to the floor (covered with calcite crust) or calcified to an older generation of secondary formations—evidence of the relative age of the event that caused the deformation.

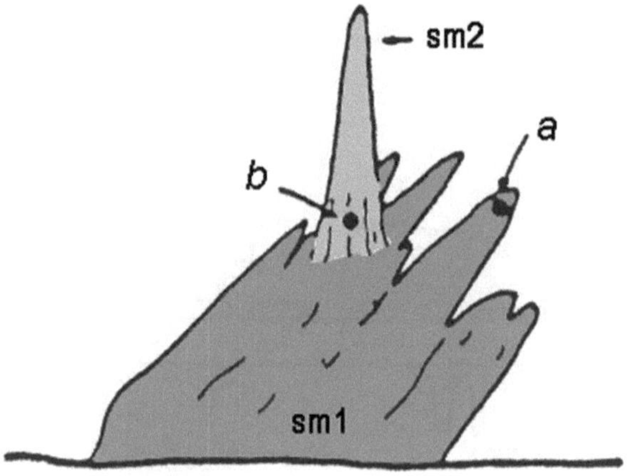

Fig. 3.6 Inclined stalagmite line (sm 1) with new vertical stalagmite (sm 2) where a and b are places for dating samples (Quinif 1997)

Samples for absolute dating to determine the time interval in which the paleoseismic event happened are selected from certain areas of deformed speleothems (Figs. 3.5 and 3.6).

3.2.2 Measurement of Natural Frequencies and Horizontal Ground Acceleration of Speleothems

Generally, the method presents the inverse approach to the morphological and statistical analyses of deformed speleothems. This complex technique is based on the hypothesis that the availability of non-broken speleothems in caves could be interpreted as absence of strong seismic events during historic times.

To our knowledge, this type of research was first applied in field studies in France and Hungary by the teams of Dr. Corinne Lacave from Résonance Ingéniours-Conseils SA (Geneve, Switzerland) and Prof. Gyozo Szeidovitz from the Geodetic and Geophysical Research Institute of Hungarian Academy of Sciences (Lacave et al. 2000, 2003, 2004; Szeidovitz et al. 2005, 2008). Important in theoretical aspect are the laboratory tests of stalagmites rupturing at the University of Liege (Cadorin et al. 2000, 2001). Theoretical modeling of the mechanical behavior of speleotems during seismic event was performed by the Saclay Mechanics and Technology Department of CEA, France (Gilli et al. 1998).

The team of Lacave explored two caves in Vercors, France: Choranche Cave and Antre de Vénus Cave. The measurement of the natural frequency was made using Polytech OFV 3001 high-resolution laser interferometer. At the top of the studied speleothem was stuck a piece of reflecting tape to ensure good reflection of the laser ray. The speleothems were excited with a light hit by rubber stick or with light breath for the soda straws. In Choranche Cave were measured 12 speleothems (stalagmites, stalactites, and soda straws) and 8 speleothems were measured in Antre de Vénus Cave—5 stalactites, 2 soda straws, and one stalactone. The study showed that most of the speleothems were not a subject of seismic motion because their natural frequencies were higher than the range of seismic excitation.

The thin poplar-like stalagmites are among the most attractive and rare speleothems. Szeidovitz et al. (2005) were the first who used especially these speleothems as paleoseismic indicators. Their studies on Baradla Cave in Hungary show that the high 5.1 m stalagmite in Olimposz Hall of the cave could be broken at a horizontal acceleration of 1.14–0.34 m/s^2. The diameter of the stalagmite is 7–10 cm and dating samples from the base and the top of the stalagmite are taken (Fig. 3.7).

Even a higher value hardly exceeds the 7 levels (1 m/s^2) of the MSK-64 intensity scale. Regarding the fact that the stalagmite is older than 100,000 years, they supposed that no strong seismic event occurred in the examined area.

The methodology of the Hungarian seismologists is based on the equation of Cadorin et al. (1998). They calculate the self-frequency of a stalagmite with the following formula:

$$f = \frac{1}{\Pi\sqrt{3ED^2/16\rho H^4}} \tag{3.1}$$

and ground acceleration resulting in fracture with the following:

$$a_g = \frac{r\sigma_u}{2\rho H^2} \tag{3.2}$$

The abbreviations are as follows: H is the length of the stalagmite (m), D is its diameter, r is its radius (m), ρ is the density of the stalagmite (kg/m^3), σ_u is the breaking tension (Pa), and E is the Young modulus (Pa). If resonance phenomena occurs as well, the

Fig. 3.7 Sampling from the top of a 5.1 m high stalagmite in Baradla Cave, Hungary (by Szeidovitz et al. 2005)

formula for the acceleration amplitude of the breaking is modified as follows: (Cadorin et al. 2001)

$$a_g = \frac{r\sigma_u}{2\rho H^2\sqrt{(1-(T/T^0))^2+4\rho^2(T/T^0)^2}} \text{ where } \rho^2 = \frac{\log^2 \varepsilon}{1,9+\log^2 \varepsilon} \quad (3.3)$$

Knowing the quenching (g) (the value of (g) can be calculated from the proportion of consecutive amplitudes of seismograms made for the determination of speleothem frequencies), D can be determined. T^0 is the self-period of the studied stalagmite.

In the frames of bilateral scientific cooperation between Bulgarian and Hungarian Academies of Sciences, four caves in Bulgaria were studied in the period 2005–2007. The results are presented in the following subchapters of the book.

Recently, this very perspective type of research was continued by Dr. Katalin Gribovszki in caves in Central Europe (Gribovszki et al. 2013a, b, c).

3.2.3 Monitoring of Recent Geodynamics

In some caves are observed indicators of recent tectonic movements on the elements of cave morphology. Such indicators are dislocated passages in the cross section of the gallery due to fault activity, tectonic mirrors, presence of tectonic breccia, spatial discordance in the couple "stalactite—stalagmite reciprocal," and dislocated or inclined stalactones (Figs. 3.8 and 3.9). These benchmarks for active tectonics in caves are described in detail in the works of Wojcik and Zwolinski (1959), Bögli (1969), Garasic (1981), Zacharov (1984), Gilli (1986), Quinif (1996), Jeannin (1990), Bini et al. (1992), Quinif (1997), Mochiutti (2004), and others.

The karstic caves are preferred as a medium for geodynamic monitoring for two reasons: (1) In most cases the tectonic deformations are fixed clearly within the karst galleries and well preserved for a relatively long period of time and (2) the cave's annual temperatures are almost constant. Together with the specific conditions of sedimentation in the caves and the absence of dynamic geomorphic processes such as excavation and weathering inherent to the Earth's surface, the underground karst forms are excellent places for installing extensometric facilities.

The robust conditions in most of the caves are a precondition to usage of mechanical micro-tectonic monitoring instruments. In several caves of Czech Republic, Slovakia, Poland, Hungary, Slovenia, and Bulgaria are installed mechano-optical extensometers, named TM-71 (Briestenský et al. 2010; Gosar et al. 2011; Kis et al. 2012; Šebela et al. 2005). The instrument was developed by Prof. Blahoslav Kostak from the Czechoslovak Academy of Sciences at the end of the 1960s (Kostak 1969) and was originally designed for monitoring of fractures and cracks in anthropogenic underground structures (tunnels, etc.). The first TM-71 extensometer in a karstic cave was installed in 1981 in Strochy pot-hole in Slovakia (Briestenský et al. 2010). Recently, more than 160 gauges were mounted all over the world, including Spitsbergen Island with its severe climatic conditions.

TM-71 works on the principle of the optical Moire effect and needs no energy source. The microdisplacements of cracks are recorded by two equal pairs of optical glass plates. The precision of the displacement recording of the device is ±0.007 mm in X, Y, and Z coordinates (Klimes et al. 2012). Detailed

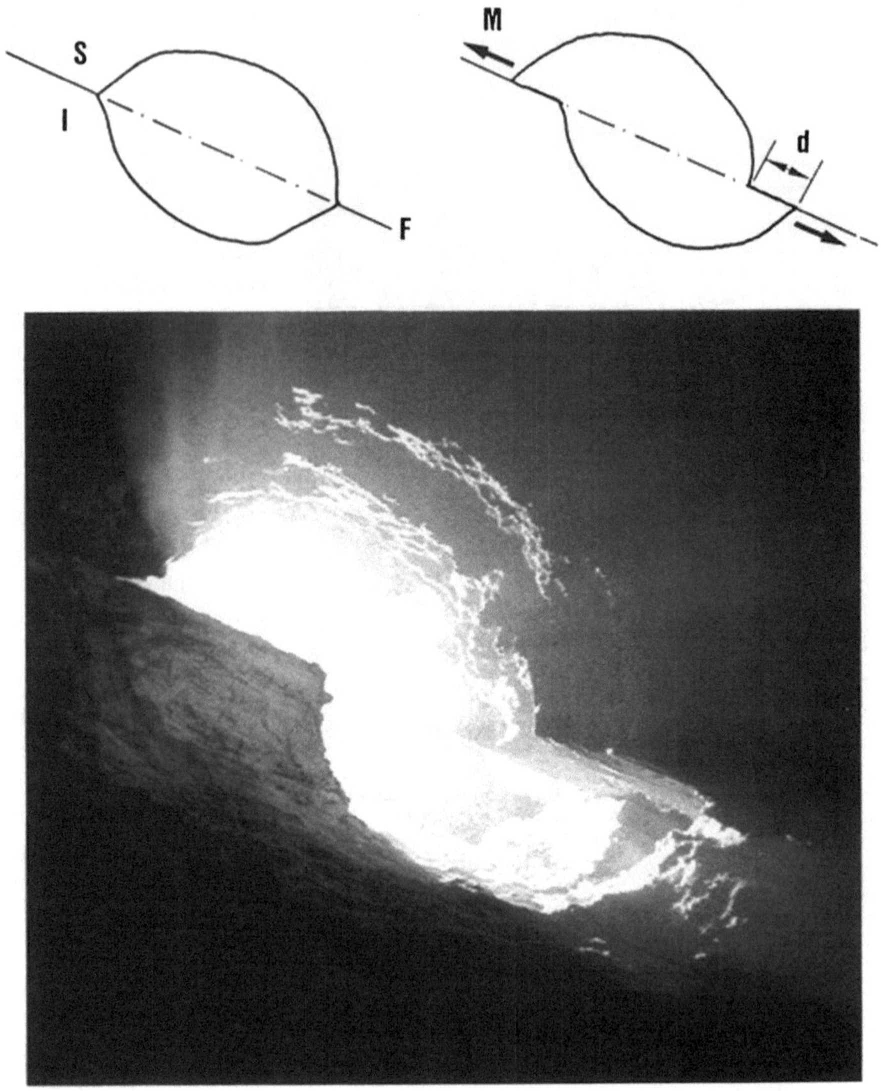

Fig. 3.8 Displaced cave gallery due to fault movement (Frasino Cave, Monte Campo dei Fiori, Italy) (Quinif 1997)

technical description of the gauge and the principles of measurement are given in Kostak (1969, 1991) and Klimes et al. (2012).

Since 1990, extensometers have been installed in six Bulgarian caves till now. The first one was in Golyamata Tsepnatina Cave (70 m long) on the Madara Plateau, NE Bulgaria. The device is a part of monitoring system of three extensometers and several marks that record the displacements within the vicinities of the UNESCO protected historical monument from seventh century Madara Horseman (the only known rock bas-relief in Europe).

In 2011, two extensometers were installed in Bacho Kiro Cave and Saeva Dupka Cave in the zone of Fore-Balkan, North Bulgaria. The installations were accomplished in a project by the Czech and Bulgarian Academy of Sciences (Fig. 3.10.).

In the frames of the Romanian–Bulgarian project MARINEGEOHAZARD dealing with geohazards on the Black Sea coast, in 2013 five extensometers were installed on the rock cliffs of the northern Bulgarian shoreline, north of the town of Kavarna. Three of them are mounted in natural karst cavities in Kaliakra Cape, Bolata Valley, and Yailata Lanslide (Fig. 3.11).

In addition to TM-71, in some caves are installed different types of extensometers. Mechano-electonical extensometers were developed in the Royal Observatory of Belgium in Brussels at the end of 1980s in the frames of cooperation with the Chinese State Seismological Bureau (Cai et al. 1989). Such devices

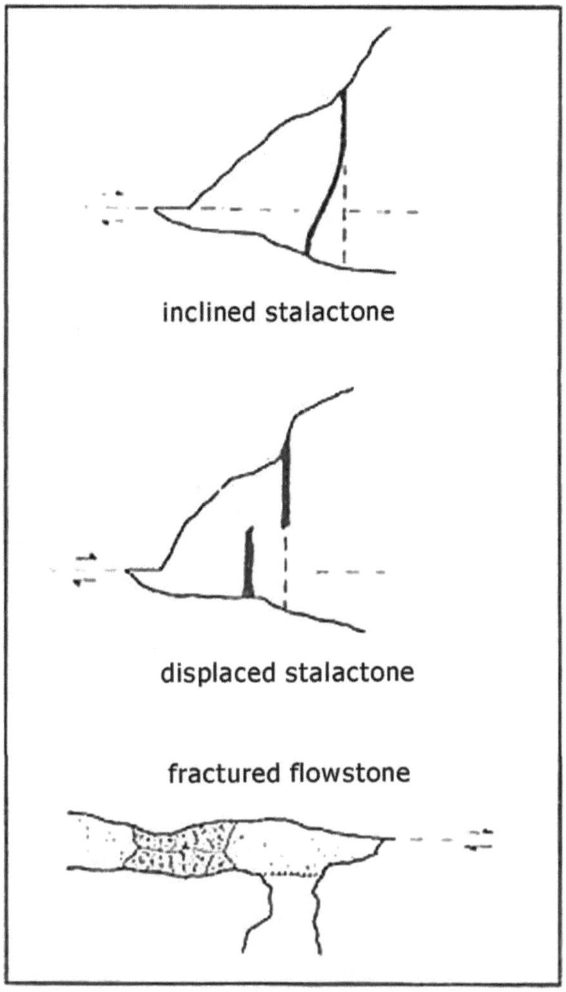

Fig. 3.9 Types of deformations of speleothems caused by active tectonics in the Siebenhengste-Hohgant cave system, Switzerland (modified after Jeannin 1990)

are installed in the underground laboratory in Rochefort Cave near Namur in Ardennes (a.k.a. Rochefort Laboratory of Geodynamics).

The show cave Grotta di Gigante near Trieste is well known because of the geodetic facilities installed by Prof. Antonio Marussi in 1966. He installed long-base pendulum tiltmeters sensitive to rotations and shear deformations of the cave. The original Marussi recording system, still in function, was photographic, very stable to humidity, and other external influences. In December 2003, a new recording system was installed based on a solid-state acquisition system intercepting a laser light reflected from a mirror mounted on the horizontal pendulum beam. The strong Sumatra-Andaman earthquake of December 26, 2004 was well recorded by the pendulums (Braitenberg et al. 2006).

3.2.4 Dating Methods

According to Sowers and Noller (1997) there are over 22 methods or groups of methods known for dating of Quaternary deposits and landforms. A classification scheme of the methods of dating of Pleistocene and Holocene was proposed by Colman et al. (1987). In this chapter is briefly mentioned the dating methods applied in tectonic studies in karst terrains. Without going beyond our competence and based on our practical experience from the karst areas in Bulgaria, we discuss some methodological aspects of the topic.

The geomorphologic dating methods are obviously based on the change in time of the geomorphological phenomena and landforms. Usually they include quantification of the geomorphic agents such as accumulation, denudation, and weathering. Based on modern quantitative values of these exogenous processes are built models that trace the development of the geomorphological landform. The obtained ages are relative (methods based on geomorphologic correlation). The ages derived by morphological methods are relative (comparative). In karst geomorphology the correlation between the cave levels and river terraces is widespread (Angelova et al. 1995, 1999; White 1988).

The most useful dating method in karst uses the natural radioactive decay of uranium. Using the U-series methods, samples from about a million years old to younger than 50 years can be dated (Sowers and Noller 1997). Recently, the U-Th isotopes were measured by thermal ionization mass spectrometry (TIMS) on 0.1 to few grams calcite samples, depending on the U content of the sample.

The deformed stalagmites and flowstones are cored and sampled for U-series dating with a drilling machine. The technique is discussed in detail in the case studies.

3.3 Case Studies

3.3.1 Studies in Stara Planina (Balkan) Mountains

The karst in Stara Planina Mountains is developed on about 20 % of the territory of this 550 km long and

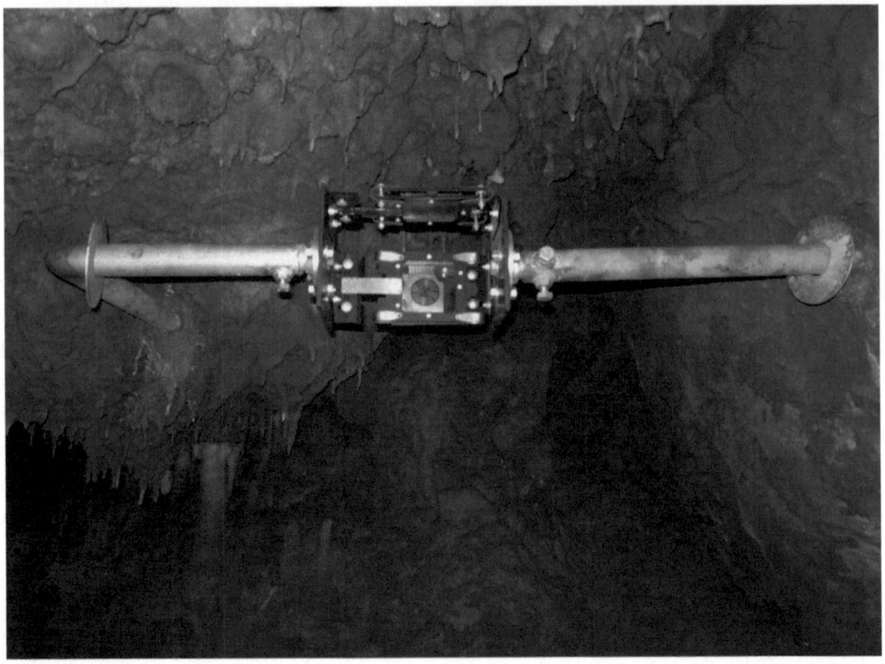

Fig. 3.10 Extensometer TM-71 in Saeva Dupka Cave, Bulgaria (*photo* Dr. N. Dobrev)

Fig. 3.11 TM-71 installed in 2013 in Domuz Maara Cave, Bolata karst valley, NE Bulgaria

2,376 m high mountain range. The karst landforms are developed mainly in Triasssic, Jurassic, and Cretaceous carbonate complexes. Due to complex geodynamic reasons, karstified rocks are in complex lithological and tectonic discordance. The karst in West Stara Planina is typical in this respect. Morphologically, the karst in Stara Planina is very diverse. The karst massifs in West Stara Planina (Chepan Mountain, Vratsa Mountain, Ponor Mountain, etc.) have typical geomorphologic features of the classical Dynaric karst: karren fields, deep gorges, blind valleys, deep pot-holes, poljes, etc. The karst geomorphology and hydrology have been studied by several researchers: Radev (1915), Angelova (2003),

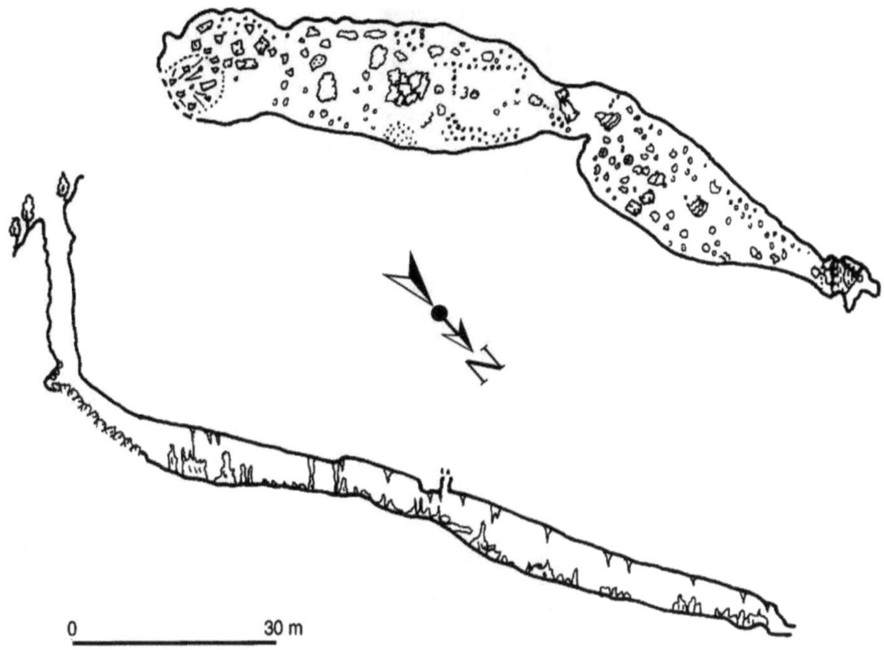

Fig. 3.12 Map and longitudinal profile of Elata Cave (after Beron et al. 2006)

Angelova and Benderev (2000), Angelova et al. (1995, 1999, 2002), Benderev (1989, 1991), Benderev et al. (2006), Kostov (1997), and others.

There are 19 cave regions in the Stara Planina Mountains according to the scheme of Popov (1970). Most of the deepest pot-holes in Bulgaria have been established here (12 caves more than 200 m deep) including the deepest Bulgarian cave Raichova Dupka Cave (−387 m).

3.3.1.1 Elata Cave

The Elata Cave is situated in Ponor Mountain (morphological unit ot West Stara Planina (Balkan) Mts.), about 55 km north of Sofia and 1.5 km SW from the village of Zimevitsa.

The cave begins with a vertical pit of 18.5 m, followed by a large gallery of length 176 m. The total denivelation of the cave is 64 m (Fig. 3.12). The cave deposits are presented by breakdowns and a number of speleothems: stalagmites, stalactites, sinter lakes (gurs), stalactones. Typical for Elata Cave are poplar-like stalagmites of height up to 3 m ("Elata" means "The fir tree" in Bulgarian). This pot-hole was discovered in 1960 by the cavers P. Beron, V. Beshkov, and T. Michev and surveyed in the following years by members of the "Akademic" Caving Club (Sofia). According to its attractive speleothems, the cave is included in the list of the Bulgarian protected natural phenomena since 1964.

Elata Cave is developed in gray Upper Jurassic limestone of the Yavoretz Formation. The chronostratigraphic range of the formation is Lower Calovian—Lower Cimeridgean. Structurally, the cave is formed along the line of a fault with SE–NW direction. The displacement of the Pleistocene and Holocene deposits on the surface can be assumed as a result of the Quaternary activity of the fault (Paskalev et al. 1992).

As a result of two strong seismic events in Ponor Mountain are established hydrological phenomena related to the discharge of the Iskrets karst springs. The springs are located 6.2 km SW of Elata Cave. During the devastating earthquake in Vrancea, Romania of March 4, 1977, the spring's discharge decreased from 5.5 to 0.5 m^3/s. After 7.5 hours there occurred a sharp increase of water levels up to 18 m^3/s, accompanied by intense water turbidity possibly connected to erosion of sediments in aqueous karst galleries and breakdowns. About that time the massif detained about 13,500 m^3 of water. A similar phenomenon was described after Svoge earthquake of March 9, 1980 (Petrov 1983; Paskalev et al. 1992).

The registered hydrological anomalies in the flow of the Iskrets springs can be explained by collapses in the karstic galleries. The catchment area of the

Fig. 3.13 Fallen stalagmite covered by new vertical stalagmite in Elata Cave

springs is about 140 km^2 and covers a significant part of Ponor Mountain. This was the reason for the search for suitable seismodeformations in karst caves above the springs.

During several visits in 2007, Elata Cave was explored by Bulgarian and Hungarian scientists: Dr. Katalin Gribovszki and Tibor Czifra from the Geodetic and Geophysical Institute of the Hungarian Academy of Sciences, Dr. Gergely Surányi from ELTE University in Budapest, and Gabriel Nikolov and Dr. Konstantin Kostov from the Geological Institute, Bulgarian Academy of Sciences. The following indications of seismotectonic activity were identified:

(A) *Fallen stalagmites*. In the cave were established 41 fallen stalagmites. Their size varies from 0.2 to 1.8 m (Figs. 3.13. and 3.14). Fallen stalagmites in some cases are calcified to the floor and covered with thick calcite crust, while others with regrowth of new vertical stalagmites on the fallen samples. Their substantial amount to their small dimensions of the cave is a sign of the impact of seismotectonic events.

The rose-diagram of the spatial orientation of the fallen stalagmites (Fig. 3.15) clearly shows the presence of a preferred direction to the north. The formation of well-defined maxima in the preferred direction of the deformed speleothems is accepted as an evidence of the seismotectonic origin of the perturbations (Dublyansky 1995, 2000; Delaby 2000). The Quaternary activation of the sub-parallel fault structure expressed on the surface near the cave is the probable cause of the seismotectonic phenomena as fallen and tilted stalagmites and the breakdown. The cave is developed in the hanging wall of the fault—an interesting relation that was confirmed in paleoseismologic research in caves from other karst regions of Bulgaria.

(B) *Inclined stalagmites and stalagmite lines*. The strongly inclined stalagmite lines in Elata Cave are a comparatively rare speleological phenomenon. The inclined stalagmites have impressive size—height up to 2 m (Fig. 3.16). Their presence can be interpreted as a result of slow subsidence of sediments below their base, but the process of inclination itself is probably helped by tectonic event.

(C) *Inclined stalactones*. In this cave were established two inclined stalactones. The first column is at an early stage of development of 55 cm height but the other is massive, with a height of 1.6 m (Fig. 3.17). These formations are an extremely rare indication for the presence of neotectonic activity and slow movement along a fault (Jeannin 1990);

(D) *Open crack in flowstones*. This fissure of length about 14 m can be observed in the boundary between undistorted, vertical stalagmites calcite crust on horizontal and inclined array of stalagmites (Fig. 3.18). The depth of the crack is 2 m.

Fig. 3.14 Fallen massive stalagmite calcified to the floor with new vertical stalagmite

(E) *Breakdown.* The gallery of the cave between 70 and 140 m is almost completely filled with blocks of large size. On the basis of morphological characteristics (presence of calcite cover or series of small stalagmites on the blocks) it can be assumed that the breakdown was approximately at Upper Pleistocene—Lower Holocene age.

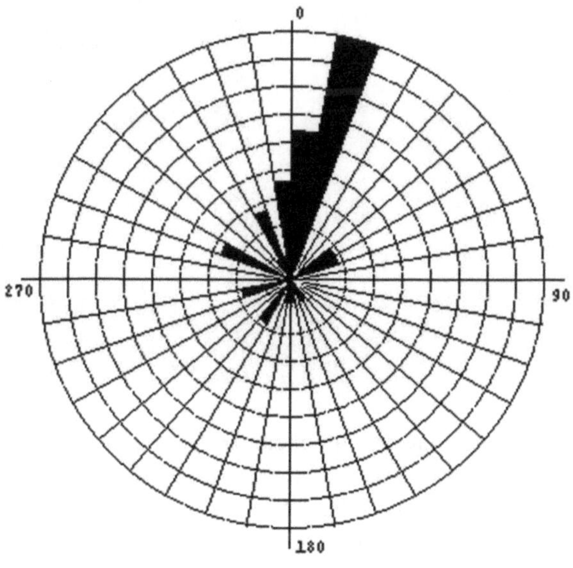

Fig. 3.15 Rose-diagram of the directions of 41 fallen and recovered with new speleothem stalagmites from Elata Cave

In the frames of bilateral scientific cooperation between Bulgarian and Hungarian Academies of Sciences, in 2007 was performed a study on the natural frequencies of stalagmites. The aim of the study was to estimate the upper limit of prehistoric ground acceleration which can break the thin and high poplar-like stalagmites in the cave.

For the purpose of the study a stalagmite with height of 237 cm and diameter between 8.3 and 13.7 cm was selected. A three geophones SM6 with natural frequency of 4.5 Hz was fixed with adhesive tape on different heights on the speleothem (Fig. 3.19).

Measuring the natural frequency of a stalagmite requires forced horizontal vibration by gentle hand hitting. The horizontal acceleration was registered by SIG SMACH SM-2 digitizer. The sampling rate of the analog-digital converter was set at 256 Hz, whereas the cutoff frequency of the anti-aliasing filter was 50 Hz. The power spectral density of the vibration was determined by Fourier transform.

The natural frequency of stalagmite is 17.8 Hz (Table 3.1.). If the natural frequency is below 20 Hz (the approximate upper limit of the frequency range of the nearby earthquakes), then resonance can occur.

Five samples of broken stalagmites in Elata Cave were collected for laboratory measurements of their mechanical properties. The samples were found on the ground at the same cave passage where the

Fig. 3.16 Inclined line of stalagmites with new generation of vertical speleothems

Fig. 3.17 Inclined massive 1.6 m high stalactone

investigated poplar stalagmite stands. The average density of the stalagmite pieces was 2.50 gr/cm^3 with standard deviation of 0.082 gr/cm^3. The Young's modulus was calculated using the data of a compressive strength test. The mean value was 7,490 MPa.

By Brazilian test the mean tensile failure stress of the samples was obtained—1.62 MPa (standard deviation of 0.42 MPa). The theoretical natural frequency and the static horizontal ground acceleration resulting in failure can be calculated using these mechanical properties (Table 3.1.).

Fig. 3.18 Open fissure in thick calcite crust in Elata Cave

Fig. 3.19 The investigated 237 cm high stalagmite in Elata Cave with attached geophones

Breaking of stalagmite with height of 237 cm can be expected for mean +1σ tensile failure stress at peak ground acceleration in the range of 350–450 cm/s^2.

Sampling and age determination were carried out. A sample from 16 cm from the top of the stalagmite was taken. The performed U/Th alpha spectrometric measurement in ELTE University in Budapest shows that the sample is not older than 6,000 years.

3.3.1.2 Razhishka Cave

The Razhishka Cave is situated in southwestern Vratsa Mountain, West Stara Planina, NW Bulgaria. The cave is located in the protected area of Lakatnik Rocks on the left slope of the Stara Planina Iskar River gorge. The big entrance at altitude of 460 m and dimensions 7 × 6 m is clearly visible from the Lakatnik railaway station.

The cave is developed in Triassic dark gray organogenic limestones from Babino Formation of the Iskar Carbonate Group (Tronkov 1981). The age of the formation is Anisian. Two major parallel reverse fault structures with direction of 70° are identified in the area (Tronkov 1965).

The Razhishka Cave has a length of 316 m and denivelation of 32 m. The cave consists of single dry

Table 3.1 The parameters of the investigated stalagmite in Elata Cave (after Paskaleva et al. 2008)

	H (cm)	$D_{Average}$ cm	Measured natural frequency, f_0 Hz	Elastic modul, E GPa	Density gr/cm^3	Tensile failure stress MPa	Theoretical natural frequency, f_0 Hz	a_g cm/s^2
Elata Cave	237	11.3	17.8	7.49	2.5 ± 0.08	1.62 ± 0.42	4.88	325.9

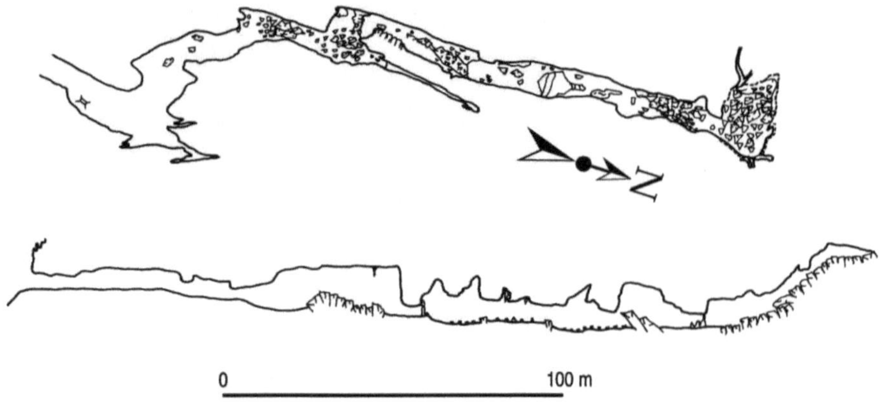

Fig. 3.20 Map and longitudinal profile of Razhishka Cave (after Beron et al. 2006)

gallery developed along a fracture with meridional direction (Fig. 3.20). The cave ends with a giant breakdown on the contact area with a transverse crack. In 1979, students at the University of Mining and Geology, M. Georgiev and Z. Andonov, carried out geophysical electrical profiling on the plateau surface above the cave to seek an extension. The measurement shows the presence of a large cave gallery after the collapse (Zlatkova 1987). Numerous attempts by cavers to find prolongation after the breakdown have been unsuccessful till now.

From geomorphological point of view, Razhishka Cave along with other caves of the Lakatnik karst basin is an emblematic example of leveled karst. The vertical cave system Razhishka Cave (the oldest upper fossil dry level)—Temnata Dupka Cave (active water cave with length of more then 6,000 m)—karst spring Jitolyub (water cave with sump) demonstrate the stages in the development of the karst process and its relation to the tectonic development of Vratsa Mountain and incising of Iskar River during the Quaternary. The vertical range of the system is approximately 220 m. Using geomorphological correlation between the relative height of the cave hypsometric levels and the river terraces of Iskar River, it can be assumed that the Razhishka Cave was formed at the beginning of the Lower Pleistocene.

During field study conducted in September 2003, in Razhishka Cave were identified the following indicators of seismotectonic activity:

(A) *Fallen stalagmites* covered with a thick layer of calcite deposits. They are distributed in the inner parts of the cave at 100–280 m distance from the entrance. The fallen stalagmites have length up to 1 m and diameter up to 30 cm;

(B) *Massive stalagmite* with height of 230 cm and diameter of 80 cm, cut by an open subhorizontal crack. The crack is filled with calcite deposits. The upper part of the stalagmite is at horizontal displacement of 7 cm. (direction 222°) and rotated to the base (Fig. 3.21);

(C) *Massive inclined stalagmite* with height of 3.5 m, cut by a crack. The inclined stalagmite is supported by its neighboring stalagmite formed near the wall of the cave gallery. Thick speleothem deposits cover the tops of the two stalagmites and form anomalous double stalactone (Fig. 3.22);

(D) *Dislocated passage* well visible on the ceiling of the cave due to vertical fault movement to the direction 10°. The dislocation is clearly traced on a horizontal distance of about 15 m, between 50 and 70 m from the entrance of the cave. The height of the cave ceiling in this part is about 4.5 m but then increases sharply and tracking the dislocation is difficult (Fig. 3.23).

Fig. 3.21 Massive stalagmite cut by an open crack. The upper part of the speleothem is rotated to its base

Fig. 3.22 Anomalous double stalactone in Razhishka Cave

In Razhishka Cave were measured the directions of 16 fallen and deformed stalagmites. The composed rose-diagram shows the existence of three maxima in the preferred direction of the deformations: 120–130°, 260–270°, and 390–400° (Fig. 3.24). The most well-defined maximum (390–400°) coincides with the orientation of the cave gallery. However, it is oriented to the main fault structure in the area—Radov Vrah reverse fault. It can be assumed that the presence of deformed stalagmites constituting this maximum is a result of the disastrous collapse, which ends the cave. The breakdown in turn, is also likely to be interpreted as a seismotectonic phenomenon of unknown age. The other two maxima are with orientation different from S–N and can be interpreted as a result of subsequent phases of intense seismic activity, depending on the behavior of the Vratsa tectonic block during the Pleistocene and Holocene. This block is characterized by intense vertical tectonic movements and active geodynamic setting.

An eventually absolute dating of the different deformed speleothems generations in Razhishka Cave will be beneficial to clarify the Quaternary tectonic history of the Western Balkan Mountains.

3.3.1.3 Varteshkata Cave

The Varteshkata Cave is a pot-hole, located in Vratsa Mountain, close to the village of Chelopek, NW Bulgaria. The vertical entrance is at 1,196 m a.s.l.

This pot-hole is 154 m long with total denivelation of 74 m and begins with 24 m vertical pit (Fig. 3.25). The cave is developed in highly karstified Upper Jurassic–Lower Cretaceous massive gray limestones from the Glozhene Formation. The Varteshkata Cave was discovered and surveyed in 1968 by cavers from Sofia Caving Club "Edelweiss" (Beron et al. 2006). The difficult access is a precondition to the conservation of the cave—till now it is rarely visited.

From a morphological point of view, the cave consists of one large gallery of height more than

Fig. 3.23 Dislocated passage on the ceiling. The amplitude of the displacement is 30 cm

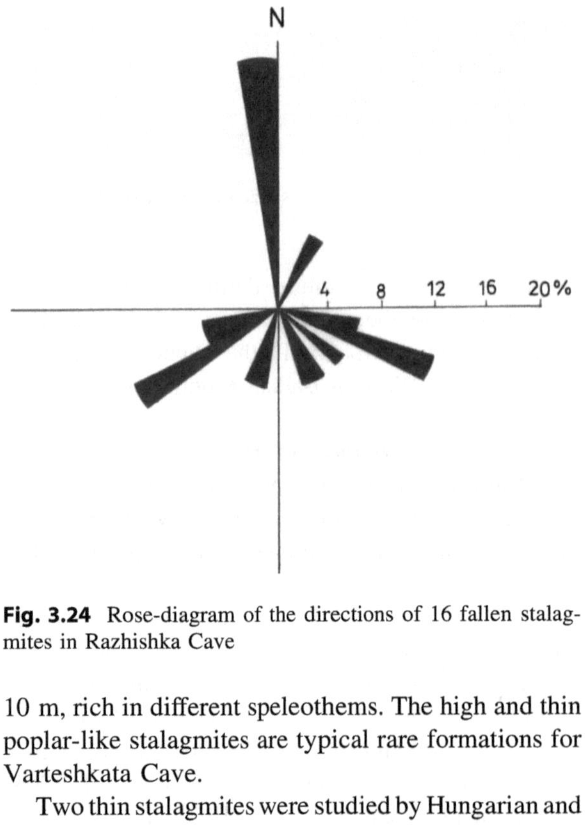

Fig. 3.24 Rose-diagram of the directions of 16 fallen stalagmites in Razhishka Cave

10 m, rich in different speleothems. The high and thin poplar-like stalagmites are typical rare formations for Varteshkata Cave.

Two thin stalagmites were studied by Hungarian and Bulgarian scientists in the Big Chamber of the cave (Paskaleva et al. 2006, 2008; Szeidovitz et al. 2008). The tallest is 365 cm high and the other is 140 cm. The used methodology is the same as that applied in Elata Cave. The diameter of the stalagmites ranges between 5.1 and 13.1 cm. The height (H) versus diameter (D) ratios are $28 \leq H/D \leq 36$ for the tall stalagmite and $23 \leq H/D \leq 28$ for the small stalagmite. The results of the dimension measurements can be seen in Table 3.2.

The results show that the value of the horizontal acceleration needed to break the stalagmites is between 0.143 and 0.438 g. These values are equal to about VIII-X degrees of the MSK-64 intensity scale.

Samples for U-series dating were taken from the top and the bottom of the 365 cm stalagmite (Fig. 3.26). The dating performed by Dr. Surányi in Budapest shows that the speleothem is rather young and its deposition is accomplished in the last 16,000 years. The age of the sample from the base of the tall stalagmite is 16,200 BP. The sample taken from the top is aged 1,100.

The results from the study can be interpreted as evidence of the absence of catastrophic seismic events during the Subatlantic phase of the Holocene in this part of Vratsa Mountain.

3.3.1.4 Labirinta Cave

The Stara Planina Iskar Gorge is situated in the Western Stara Planina Mountains (Western Balkan) and is one of the remarkable geomorphological phenomena in Bulgaria. The total length of the gorge is 67 km. Labirinta Cave is located in the eastern marginal part of the gorge, near Cherepish railway station

Fig. 3.25 Longitudinal profile and map of Varteshkata Cave (after Beron et al. 2006)

Table 3.2 Parameters of the investigated stalagmites in Varteshkata Cave (after Paskaleva et al. 2008)

	H (cm)	$D_{Average}$ cm	Measured natural frequency, f_0 Hz	Elastic modul, E GPa	Density gr/cm^3	Tensile failure stress MPa	Theoretical natural frequency, f_0 Hz	a_g cm/s^2
Tall stalagmite	365	11.8	4.5	7.49	2.5 ± 0.08	1.62 ± 0.42	2.15	143.5
Small stalagmite	140	5.3	10.6	7.49	2.5 ± 0.08	1.62 ± 0.42	6.56	438.1

(Fig. 2.2.6. in Chap. 2). The oval-shaped entrance is situated on the right side of the Iskar river, at an elevation of nearly 105 m above the base level. Some data about the cave are reported by Ilieva et al. (1981) and Kostov (1999) (Fig. 3.27).

The cave is developed in the massive organogenic limestones of Aptian age (Bedulian) of the Cherepish formation, Vratsa Urgonian group (Nikolov et al. 1972). These carbonates are highly karstified—for example, in the Cherepish karst area more than 130 caves are currently surveyed within a 4 km^2 area. From a tectonical point of view, the cave is situated in the SE wedge-shaped part of the Zgorigrad anticline (Jordanov et al. 1961). The latter occupies most part of the Vratsa Mountain and according to the classical concept for the tectonic structure of Bulgaria (Yovcev 1971), it is a structural unit belonging to the Stara Planina Dislocation Strip.

The morphology of the studied cavity includes a series of parallel passages striking 140–150 g developed along a fault and connected via difficult narrow transversal passages (Fig. 3.2). The maze-like cave patterns (the network type of cave—sensu Palmer 1984) are an indicator of a distinct structural control on the speleogenesis.

The total length of the cave is 255 m and its vertical extent ranges over 15 m. In some places the original morphology is concealed by collapse phenomena—gravitational deposits (breakdown blocks of various sizes). Speleothems were also identified—dry and weathered dendrites which cover large parts of the cave walls. The survey was completed on December 6, 1977 by members of the "Academic" speleoclub Sofia (Fig. 3.28).

According to the results of former investigations (Ilieva et al. 1981; Angelova et al. 1995; Kostov 1999, 2008), based on methods of relative geomorphology (correlation between caves levels and terraces of the Iskar River) it has been inferred that the relative age of Labirinta Cave is Early Pleistocene.

From the comparatively wide range of neotectonic speleoindicators described in previous works

Fig. 3.26 Sampling of the tall stalagmite in Varteshkata Cave

(Jeannin 1990; Bini et al. 1992; Dublyansky 1995) the following elements are used in the present report:

(A) *Displaced (shifted) sections in the cave passages*—the original cross section of the passage has been altered as a result of the action of a certain fault.

(B) *Speleothems which cover the displacements* and in other cases are fractured by them.

Over the period March–June 1997, several visits to Labirinta Cave led to the following observations:

I. *A subvertical displacement* with an amplitude of 16–17 cm, along a normal fault, excellently preserved in a primary cave formation—rock bridge or "pillar" structure—sensu Jeannin (1990) (Fig. 3.29). In this passage the fault-plane strikes 146 g with 808 NE dip. This fault can also be traced in the cave over nearly 50 m but in other cross sections of the passage, the shift is not so distinctly visible because of the existing gravitational and weathering processes.

The occurrence of a "hanging" dendritic cover over the fault-plane implies the existence of two phases of movement. The precipitation and lithification of the speleothems occurred during a relatively calm period between the two phases of tectonic activity (Fig. 3). In terms of morphology, the first

Fig. 3.27 The entrance of Labirinta Cave

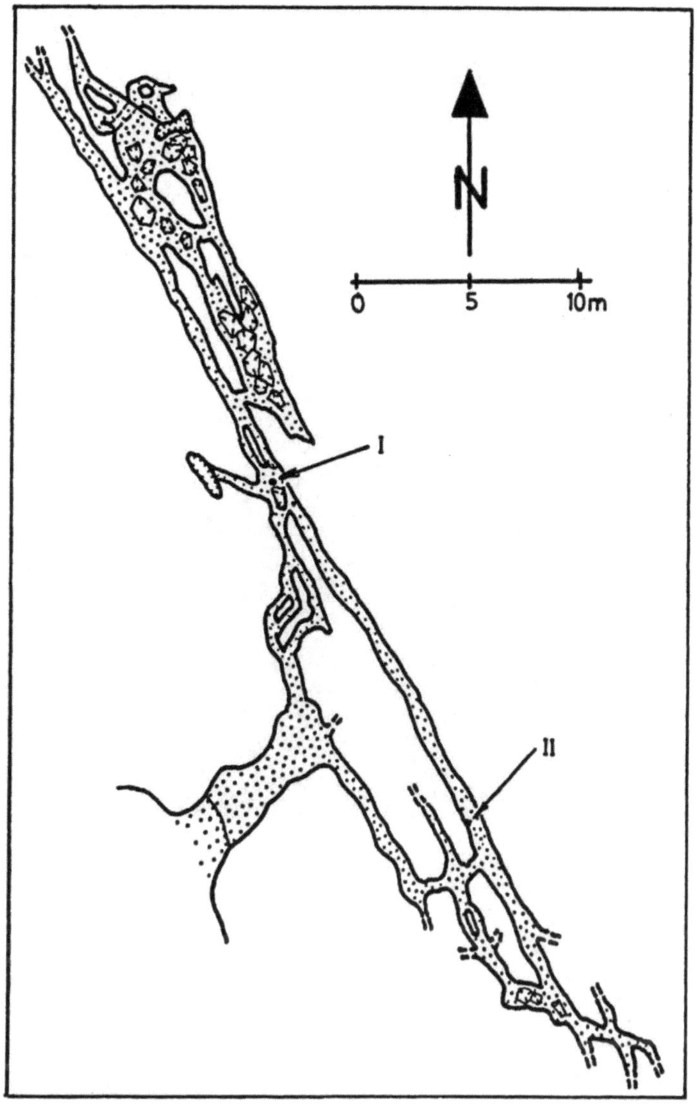

Fig. 3.28 Map of Labirinta Cave. I and II: sites of the measurements

phase that ranges from the genesis of the cave (Early Pleistocene) to the period of the precipitation of the sinter is more distinct (12–14 cm displacement amplitude). The second phase is recognized in a 2–3 cm long displacement. Unfortunately, the absence of speleothems dating renders the above conclusions debatable.

II. A subhorizontal displacement striking 160 g and dipping 25 g to the SW is situated on the wall of the same passage (Fig. 3.30). The shift amplitude is 10–14 cm, but there are some additional 8–10 cm which could be taken into account, located between the overhang and the foot of the walls. The fracture broke the sinter deposits (dendrites). Taking into account the absence of these speleothems on the fault-plane, it can be inferred that this fracture is relatively younger than that described above. It is also possible that this crack has a local cause—gravity or sliding along a slope.

Based on the performed investigations, it has been established that Labirinta Cave is developed in an area of intense Quaternary tectonic activity. According to the intensity scale of unstable karst phenomena of Gilli (1995), this cave ranges in the 4th degree: decimetric to decametric displacement of passage sections. In previous karstological investigations (Ilieva et al. 1981, Angelova et al. 1995) mostly vertical neotectonic movements and their importance

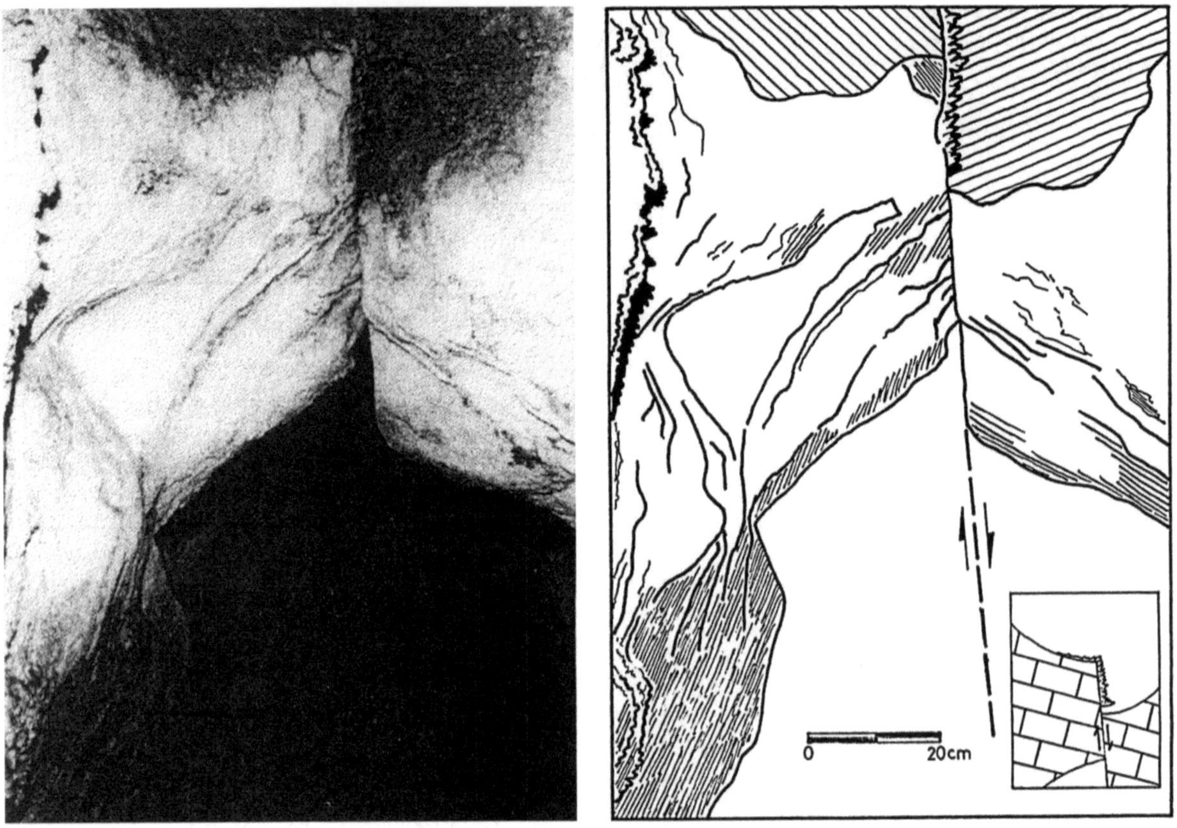

Fig. 3.29 *Left* Detail of a passage displaced by a fault. *Right* Explanation of the same passage shifted 17 cm by a normal fault. The occurrence of "hanging" speleothems on the fault-plane implies two movement phases

Fig. 3.30 Displaced passage in Labirinta Cave

in the development of different stages of typical step-like karst within the boundaries of the study area have been considered.

The post Upper Pleistocene deformations examined here, distinctly visible in the cave passages, can be accepted as evidence in support of the thesis of Tzankov and Nikolov (1996) about the occurrence of a regional extensional tectonic stress field of NNE–SSW orientation during the Neogene–Quaternary structural stage in NW Bulgaria. The subhorizontal extension environment is a precondition to the observed type of faulting, i.e., prevalence of normal and listric faults with their dip parallel to the stress direction. The extensional stress field possibly influenced the initiation of the Labirinta Cave configuration and also controlled the main directions of the cave passages in the surrounding part of the Zgorigrad anticline—N–S and NW–SE prevalently (Kostov 1997).

3.3.1.5 Saeva Dupka Cave

Saeva Dupka Cave is located in the Middle Fore-Balkan, near village Brestnitsa, Lovech District. The entrance is found on the south slope of the small Bretsnishko polje with an area of 9 km^2. The altitude of the cave is 510 m. The entrance with dimensions of 4.2 × 4.5 m is opened at the base of a small rock cliff.

The Saeva Dupka Cave is formed in the massive light Titonian limestones of the Brestnitsa Formation. The age of formation is Berriasian. The considerable thickness of the limestones from Brestnitsa Formation, the low-angle slopes, and the high tectonic fracturing are favorable conditions for intensive karstification. The karst process is supported by faults, expressed morphologically on the surface. One of them is dipping to the north normal fault with parallel direction, visible along the southern slopes of the Brestnitsa polje, close to Saeva Dupka Cave. This fault is marked with the presence of the impressive abysses Big Ledenitsa Cave (−55 m), Small Ledenitsa Cave (−40 m), and the pot-hole Partizanska Cave (−107 m, with 98 m vertical pit).

The first data for Saeva Dupka Cave were published at the end of the nineteenth century by Herman and Karel Shkorpil in their monograph on the karst waters in Bulgaria (Skorpil and Skorpil 1898).

From a morphological point of view, Saeva Dupka Cave consists of one large gallery, which separates five big chambers (Fig. 3.31). In the first two halls the cave is developed along a sub-parallel fault line. The total length of the galleries is 210 m and the denivelation of the cave is 40 m.

During field studies conducted in the period 2000–2005 in Saeva Dupka Cave the following indices of seismotectonic activity are described:

(A) *Fallen stalagmites*. In the cave were found 16 fallen stalagmites covered with thick calcite sediments. The overlay covering deformations are dry, without recent dripping, suggesting the considerable age of the deformations (Fig. 3.32);

(B) *Dislocated stalactone* with diameter of 35 cm (Fig. 3.33). A displacement with an amplitude of more than 40 cm in south direction is measured.

(C) *Horizontal crack in a massive stalactone* (Fig. 3.34). The stalactone has a diameter of more than 3 m. Through the crack is observed a minimal displacement in S–SE direction.

(D) *Breakdown*. The breakdown in Saeva Dupka Cave is one of the biggest in the Bulgarian caves. The fallen blocks cover an old generation of speleothems that can be found in a few places. Some of the boulders reach up to 3 m and are welded to each other. In some cases they are covered with new speleothems. Popov connected the brekadown occurrence and the formation of the eponymous chamber (The Breakdown Chamber) as a result of catastrophic paleoearthquake (Popov 1979).

The built rose-diagram of the orientation of the fallen stalagmites in the cave shows clearly the preferred direction to the south (Fig. 3.35). The three distinct peaks can be combined in common interval of 170–220 g. This fact can be interpreted as a result of the activation of the fault south of the cave during the Pleistocene. One may suppose that the formation of the breakdown in the cave is supported by this phase of intense tectonic activity.

In 2012, in Sueva Dupka Cave was installed TM-71 extensometer in the frames of bilateral scientific cooperation between the Czech and Bulgarian Academies of Sciences.

3.3.1.6 Troana Cave

The Troana Cave is situated in the central part of the Belyakovo Plateau which is a morphographic unit of the Central Fore-Balkan, North Bulgaria. The small precipitate entrance is located about 3 km west from the village of Emen. The cave is developed in highly

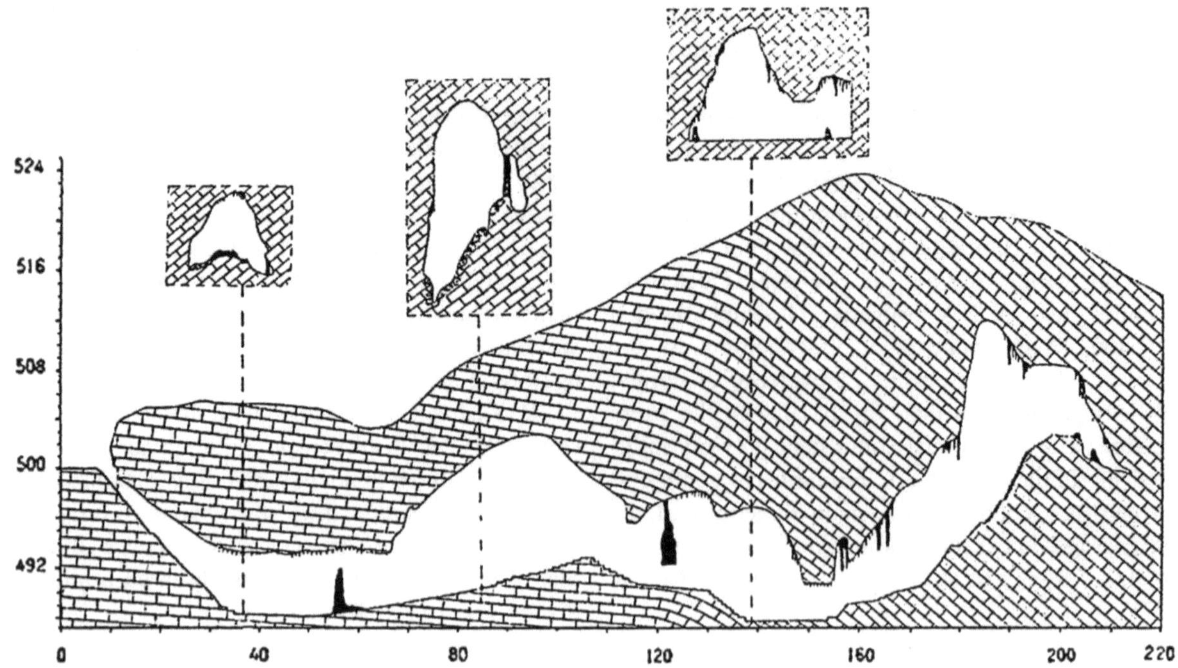

Fig. 3.31 Longitudinal profile and cross sections of the gallery of Saeva Dupka Cave (after Popov 1979)

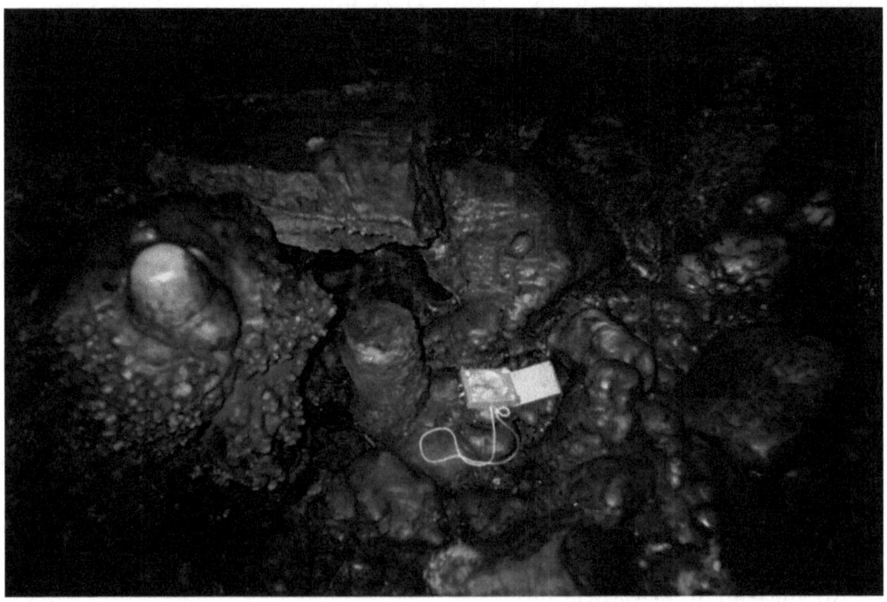

Fig. 3.32 Fallen stalagmites recovered with thick calcite deposits and new speleothems in Saeva Dupka Cave

karstified limestones of Upper Barremian–Lower Aptian age from the Emen Limestone Formation, Lovech Urgonian Group (Hrischev 1966). The maximum thickness of the stratigraphically complicated Emen Formation is about 400 m with dip up to 8°–10° to N–NW (Hrischev and Nedialkova 1992) (Fig. 3.36).

From a seismological point of view, the Troana Cave is located in one of the seismically strongest active regions in Bulgaria. During the last century,

Fig. 3.33 Displaced stalactone in Saeva Dupka Cave

three strong seismic events took place here—the Gorna Oryahovitsa earthquakes of June 14, 1913 (Ms = 7.0, Focal depth = 7 km) and December 19, 1914 (Ms = 4.7), and the Strazhitsa earthquake of 1986 (Ms = 5.7). According to historical data, nine earthquakes with magnitude up to 7.2 are known in the period 536–1890. The epicenters of these events are Veliko Tarnovo, Gabrovo, Lyaskovets, and the village of Arbanasi (Grigorova and Grigorov 1964, Orizova-Stanishkova et al. 1996).

The Troana Cave is mainly made up of a large downdipping gallery with a stream flowing along its length and a maze system of fossil passages which together form a total length of 2,750 m (Fig. 3.37). This ponor cave is entered from a doline which collects the surface streams (only during peak runoff) of two dry valleys. The entrance parts are narrow with few steps up to 3 m deep. After the entrance squeezes, the original morphology of the cave is blurred by collapse phenomena—the floor is covered by impressive breakdowns, partially coated by speleothems.

The mezokarstic deposits are presented by fluvial, detritic ones and plenty of various attractive speleothems: stalactites, stalagmites, columns, flowstones, draperies, etc. According to its rich sinter decorations,

Fig. 3.34 Fractured massive stalactone in Saeva Dupka Cave

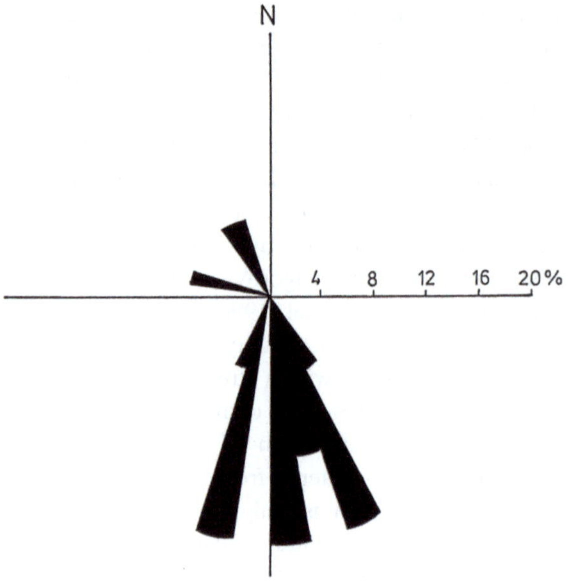

Fig. 3.35 Rose-diagram of the directions of 16 fallen and covered with new speleothems stalagmites

the Troana Cave is one of the remarkable caves in Bulgaria. The fluvial sediment accumulations in the main trunk of the cave seem to be several meters in thickness—an evidence about the high mass transport energy of the underground stream.

The cave was discovered and explored at the end of the 1980s by cavers from "Prista" speleoclub—Rousse. Until now, it is relatively well protected against human impact and in this sense represents a good object for palaeoseismological studies. In spite of this, during the field studies in Troana, we particularly explored places that were relatively difficult to reach: niches, old levels of the stream, etc.

Our search for deformations with possible coseismic origin was in the parts up to the first sump of the Troana Cave. From the comparatively wide range of palaeoseismic speleoindicators, the following elements were established there (Kostov et al. 2000):

(A) *Fallen stalagmites*. The fallen stalagmites are the most expressive sismotems. The "stalagmite's cemeteries" (sensu Quinif 2000) in some caves represent the best evidence for palaeoseismical activity. Unfortunately, their quantity in the Troana Cave did not permit us to accomplish a statistical study of the preferred orientation of the broken fragments. Only a few fallen stalagmites (secondarily recovered with calcite) have been established. The average length of the stalagmites was 10–20 cm;

(B) *Displaced stalactones*. One of the attractive features of the studied cave are the displaced columns. There are observed decametric displacements

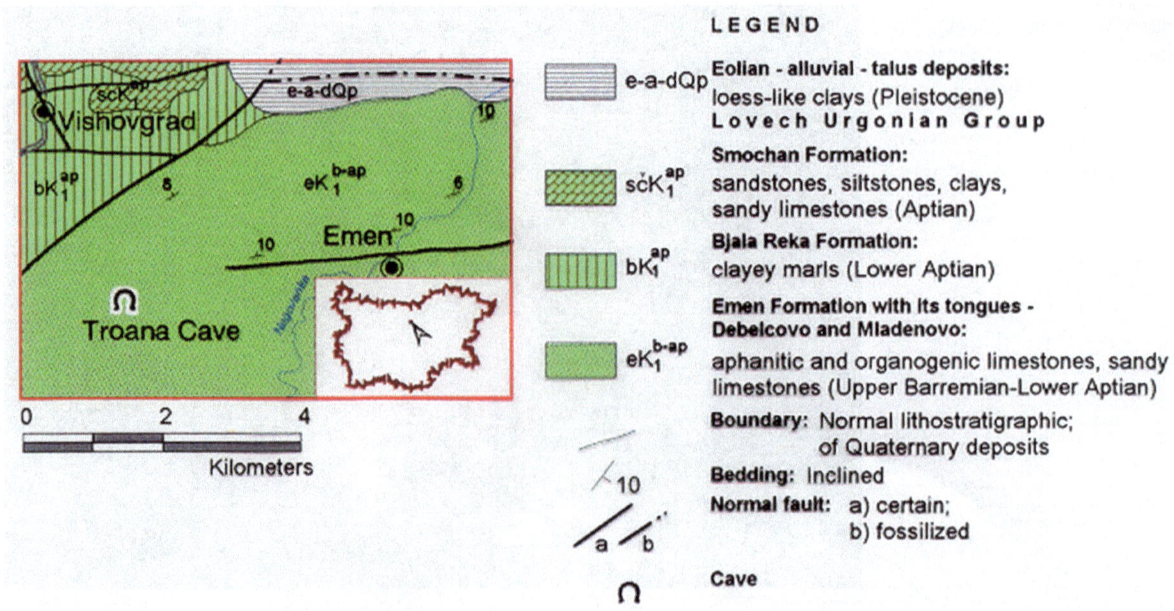

Fig. 3.36 Location of Troana Cave and geological map of the area based on sheet "Sevlievo" from the Geological Map of Bulgaria in scale 1:100 000 (Hrischev and Nedialkova 1992)

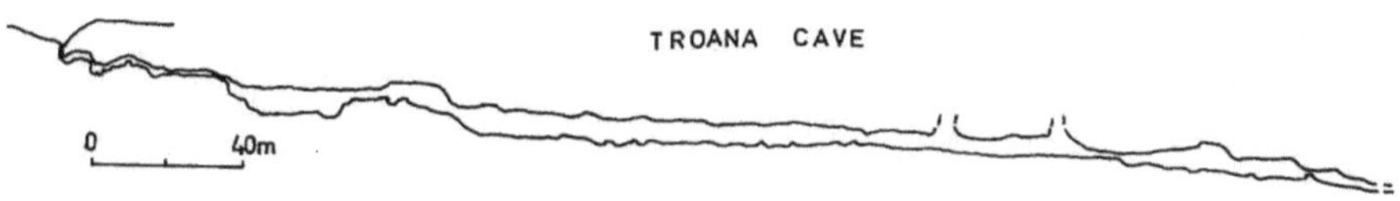

Fig. 3.37 Longitudinal profile of the galleries up to the first sump of Troana Cave

Fig. 3.38 Displaced massive stalactone in Troana Cave

of few stalactones with amplitude of 45–50 cm (Fig. 3.38);

(C) *Broken sodastrows*. Being the initial stage in the stalactites development, the soda straws are speleothems that are possibly the most amenable to the coseismic effects. Gilli and others reported that after the February 18, 1996 earthquake in the French Pyrenees (Ms = 5.2), the only effect observed in caves were breaking of soda straws (Gilli et al. 1999).

In some places of the Troana Cave are established broken soda straws. The observed stalactite fragments are of maximum length of 20 cm. The studied speleothems are now soldered with calcite onto the floor (Fig. 3.39).

The observed deformations in Troana Cave are of great interest for research on the historical seismicity of the region. Detailed studies of other caves in the area in the near future will possibly collect new data for tectonic deformations.

3.3.1.7 Golyamata Tsepnatina Cave

Golyamata Tsepnatina Cave is located on the Madara Plateau near the town of Shumen, NE Bulgaria. The cave is located on the NW periphery of the plateau, 80 m above the Madara Horseman—the only known rock bas-relief in Europe. This extraordinary historical monument is dated from the early period of the Bulgarian's state development (1-st Bulgarian kingdom) and was carved about 1,200 years ago. Madara Horseman is included in the list of the World cultural heritage of UNESCO (Fig. 3.40).

The cave entrance is accessible by 25 m descent from the edge of the plateau using single rope technique where a concrete covering, built for conservation purposes is reached. After the artificial entrance and another 7 m rope descent, the main gallery of the cave can be reached. Developed in Cenomanian sandy limestones and limy sandstones, the cave is a good example for structural control in karst.

The cave is 70 m long and formed along a fracture with direction NW–SE. It is dry and unbranched. The cave sediments are presented by weathering detritus, breakdowns, and organogenic material (guano) (Fig. 3.41).

The cave was investigated and surveyed in 1971 by cavers from "Galata" Caving Club—Varna. A second

Fig. 3.39 Broken and calcified to the floor soda straws in Troana Cave

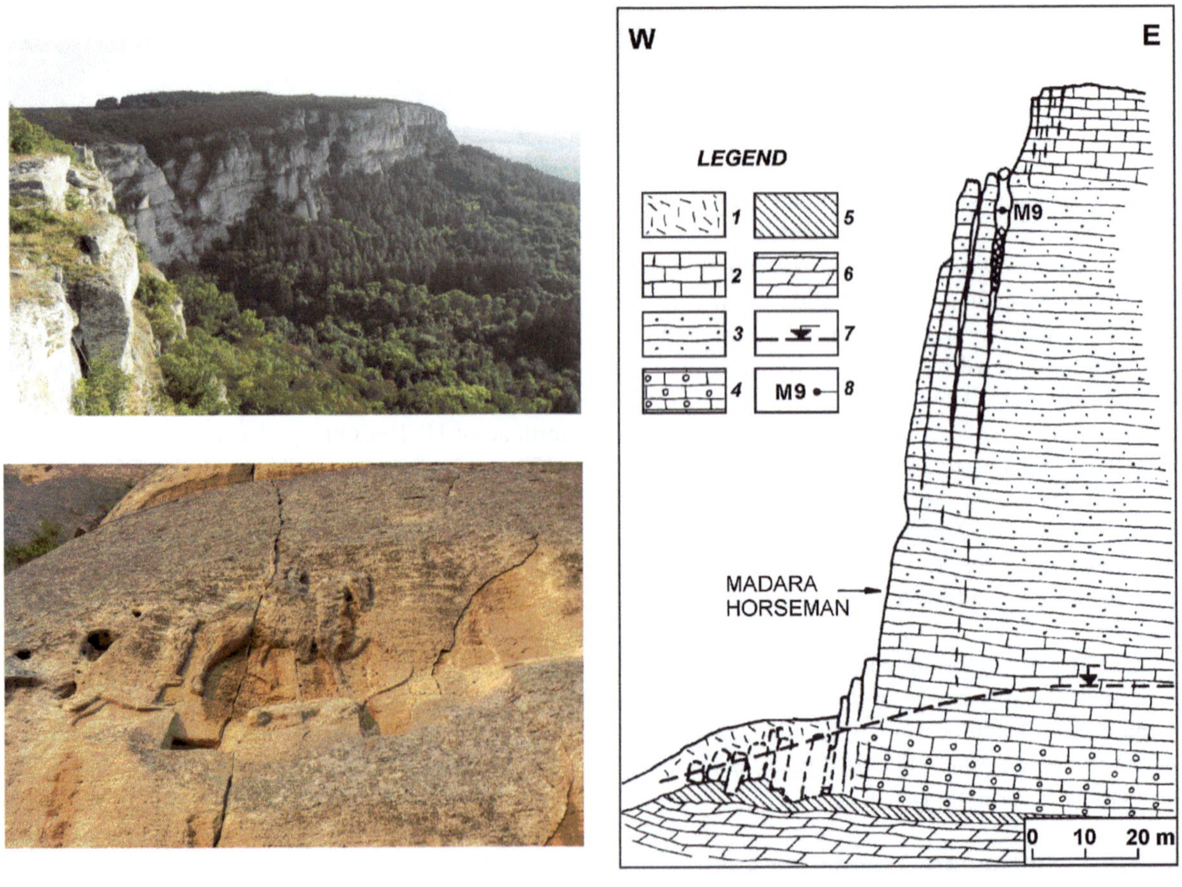

Fig. 3.40 *Left* View on the Madara Plateau above the cave (*top*) and UNESCO monument Madara Horseman rock bas-relief (*bottom*). *Right* Engineering Geological profile through the Madara plateau (after Frangov et al. 1992; Kostak et al. 1998): *1*—Deluvium; *2*—Sandy limestones; *3*—Calcareous sandstones; *4*—Conglomeratic limestones; *5*—Clay; *6*—Marls; *7*—Groundwater table; *8*—Extensometer in Golyamata Tsepnatina Cave

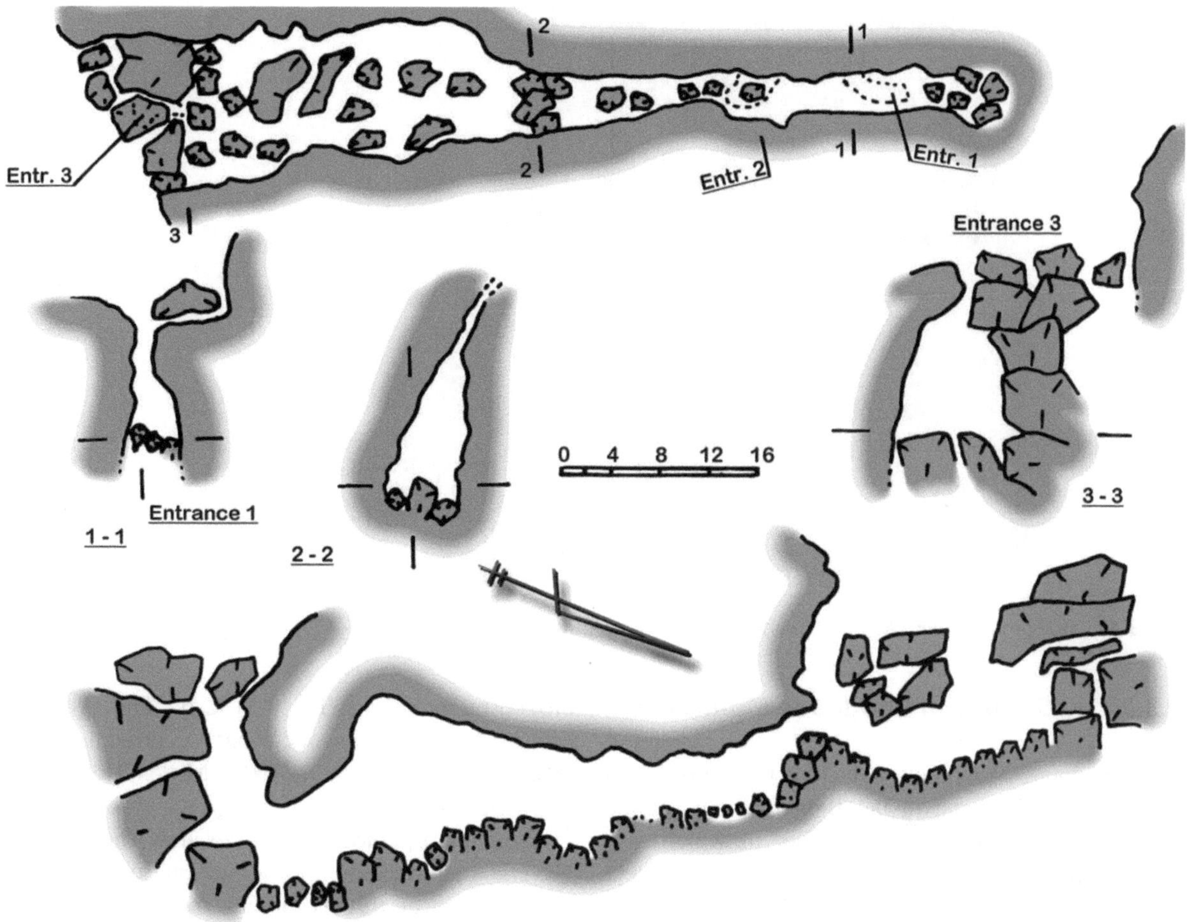

Fig. 3.41 Map, cross sections, and longitudinal profile of Golyamata Tsepnatina Cave

detailed survey was made in 1985 by Shumen cavers A. Spasov and M. Mirchev ("Madarski Konnik" Caving Club). Description of the cave was made in Mircheva et al. (2004). The cave has no special protection status but is situated within the boundaries of the National Historic-Archaeological Reserve "Madara".

As mentioned above, an extensometer TM-71 was installed in the cave in April 1990, which was a part of the monitoring system for study of the recent microdisplacements around the monument together with other two gauges TM-71 near the Madara Horseman bas-relief and a few pin marks. The data from the instruments are monitored by the Geological Institute of Bulgarian Academy of Sciences several times during the year (Fig. 3.42).

Until 1994, these movements were characterized by a pronounced sinusoidal curve showing the trend and some sharp movements with various origin (Dobrev and Avramova-Tacheva 1997). The monitoring data from the last 10 years are discussed in a special publication in preparation.

The response of the TM-71 instrument to the catastrophic earthquake in Izmit, Turkey on August 17, 1999 ($M = 7.4$) is interesting. The epicenter was located 370 km S–SE from Madara Plateau. This earthquake was felt in many parts of Bulgaria, including the nearest towns of Shumen and Varna. Immediately after the earthquake measurement was carried out. It was found that the extensometer been damaged due to strong fluctuation motion (shaking) of the two slices. After reconstruction of the gauge, we made accurate reconstruction of the movement taking place during this earthquake. It was found that: $\Delta X = +6.91$ mm shortening of the fissure; $\Delta Y = +46.78$ mm horizontal slip to S-SE; $\Delta Z = +\ 10.43$ mm vertical movement (uprising) of the rock slice.

Fig. 3.42 Monitoring of the TM-71 in Golyamata Tsepnatina Cave

3.3.2 Studies in Rhodopes Mountains

The Rhodopes Mountains stretch between Bulgaria and Greece. The karst in the Bulgarian part of the massif (Rila–Rhodopean karst zone, to which also belong Pirin and Slaviyanka mountains) is present almost everywhere and occupies 1,966 km^2. It is formed mainly of Proterozoic marbles from the Dobrostan Marble Formation. More than 600 caves have been surveyed in Rhodopes till now and the longest is Yagodina Cave—8,501 m. The deepest potholes are Banski Suhodol 9/11 Cave in Pirin Mountain (more than 300 m according to the results from the Summer Caving Expedition "Banski Suhodol 2013") and Drangaleshka Dupka Cave with depth of 255 m.

3.3.2.1 Lepenitsa Cave

The Lepenitsa Cave is a spring cave situated in the northern part of the Batak Mountain which forms a part of the Western Rhodopes Mts. The small entrance is located on the left bank of Chuckura River, nearly 13 km south of the town of Velingrad. It is formed of highly fractured Proterozoic marbles of the Dobrostan formation (Fig. 3.43).

The cave is located in one of the seismically strongest active regions in Bulgaria. The maximum observed shakeability in terms of intensity is determined as I = 7 MSC-64 and for a period of 10,000 years, the maximum prognostic intensity is I = 9 MSC-64. During the last century, two strong seismic events occurred there—the Velingrad earthquakes in 1905 (M = 4.2) and 1977 (M = 5.3) (Fig. 3.44).

From a morphological point of view, the Lepenitsa Cave is a good example of a multiphase cave system that consists of three different stages—active level ("The water gallery") and two fossil upper levels. The height difference between the levels does not exceed 30 m. The total length of the cave passages is 1,525 m. The plan of the cave was made in the period 1970–1972 by the speleologist Petar Tranteev with the help of Konstantin Spasov, Martin Tranteev, and Boyan Tranteev. The main gallery has a length of about 800 m (Fig. 3.45).

The cave deposits are fluvial, detritic, and numerous speleothems: stalactites and stalagmites of different types, columns, draperies, gurs, etc. The wide occurrence of fractured, broken, and inclines speleothems with possible coseismic origin was reported by Shanov et al. (1998), Kostov (2000, 2008).

The prevalence of cracked, broken, or inclined speleothems with possible coseismic origin was first reported by Shanov et al. (1998), (2001), and Kostov (2000). During several field studies conducted in the period 1998–2005, in the cave are identified the following indicators of seismotectonic effects on cave sediments:

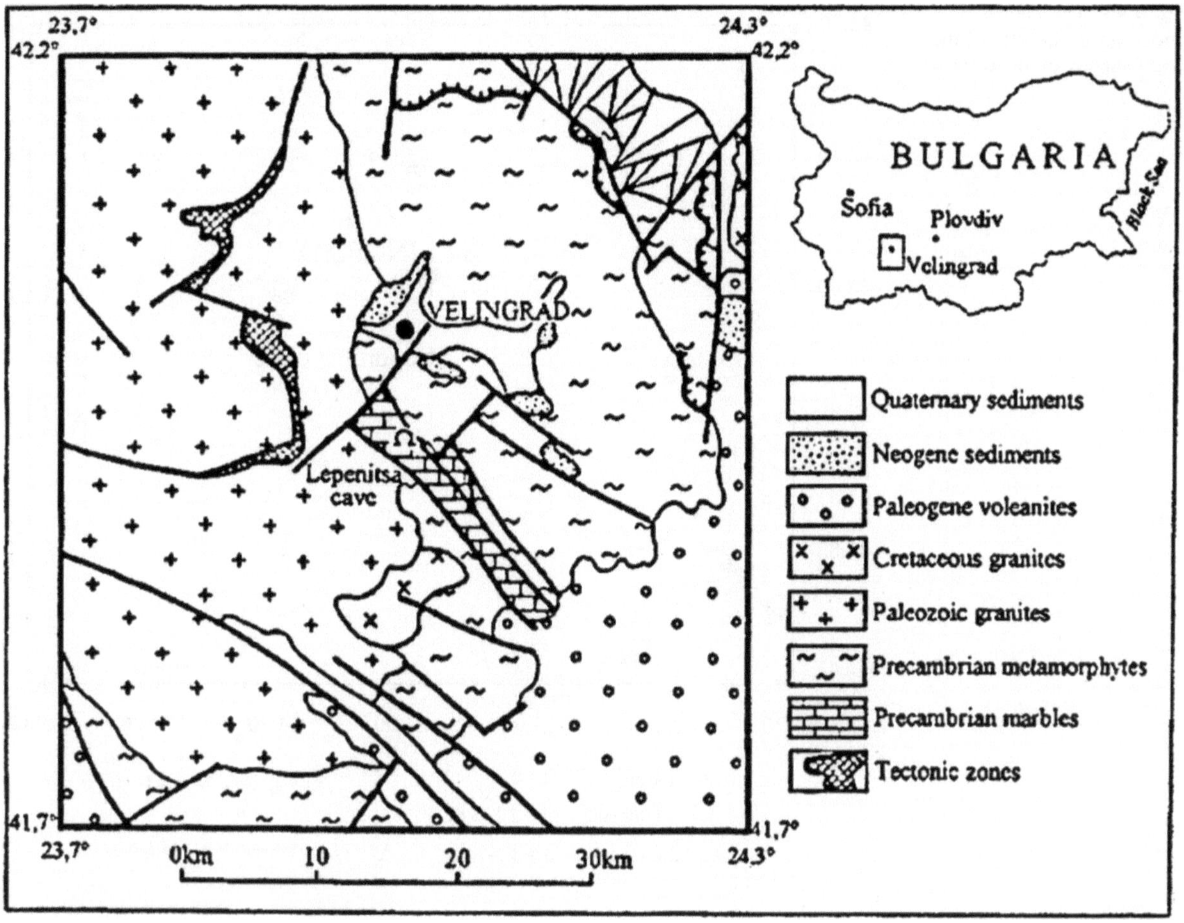

Fig. 3.43 Location of Lepenitsa Cave and geological map of the area (after Shanov et al. 2001)

(A) *Fallen and calcified to the floor stalagmites*. The perturbed stalagmites are an average length of 40 cm with secondary overlay of thin sinter deposits—evidence of the relative age of the deformation. In some cases, on the horizontal fallen fragments was observed the formation of small active stalagmites. The described samples fell on flat, horizontal floor, which excludes their secondary predisponation after failure.

(B) *Inclined massive stalagmite lines* covered with a new generation of vertical stalagmites (Fig. 3.46). The inclined stalagmite series are up to 2.5 m high. This seismotectonic phenomenon is formed in the dry section of the cave without presence of fluvial sediments under the stalagmites and no signs of their existence in the past. The subsidence of sediments is excluded as a possible cause of inclination.

(C) Open fissures in massive stalagmites (Fig. 3.47).

The detailed geomorphological study of the cave and the statistical analysis of the geometry and orientation of the deformed speleothems may prove the coseismic origin of the deformations. In the first fossil level ("The main gallery") of Lepenitsa Cave, the directions of 68 broken stalagmites or stalagmite fragments were measured and a rose-diagram was made. Most of the studied stalagmites were candle shaped of average length of 40 cm and secondarily recovered with thick sinter deposits. In some cases, recent small active stalagmites developed upon the broken ones. The selected samples have fallen down on flat and stable horizontal floor, which excludes their secondary disposition.

The diagram shows clearly expressed preferred orientation (Fig. 3.48). The maxima (360–370 g and 390–400 g) generally correspond to a direction pointing to the town of Velingrad—the epicenter of

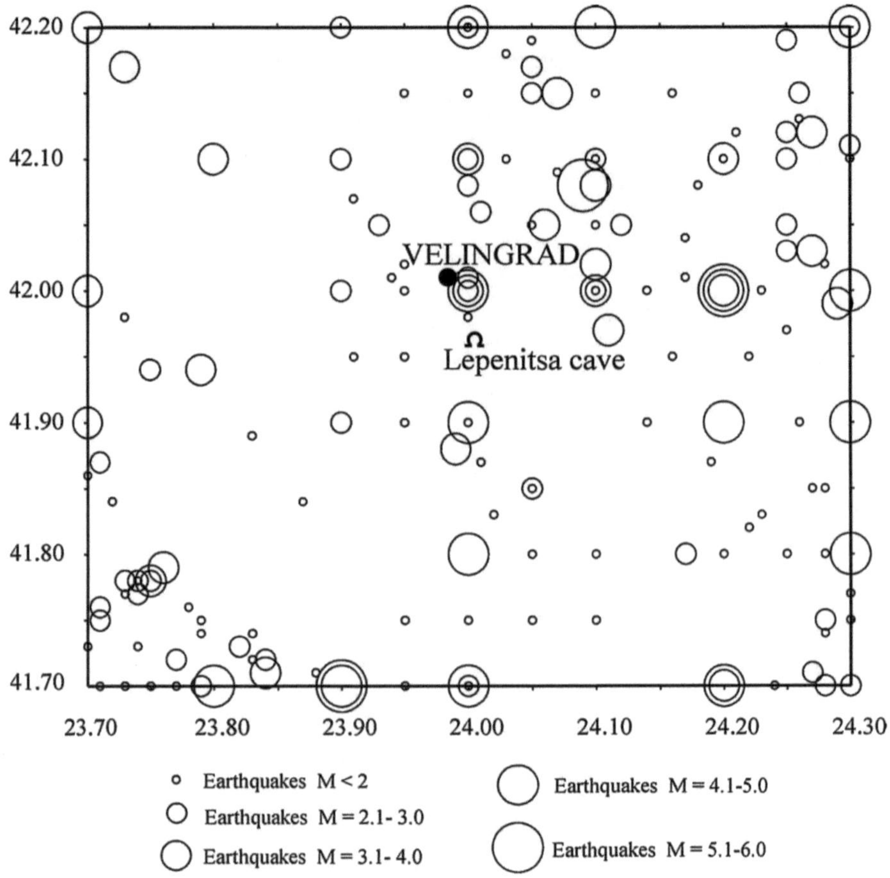

Fig. 3.44 The known epicenters in the area of the cave (Shanov et al. 2001)

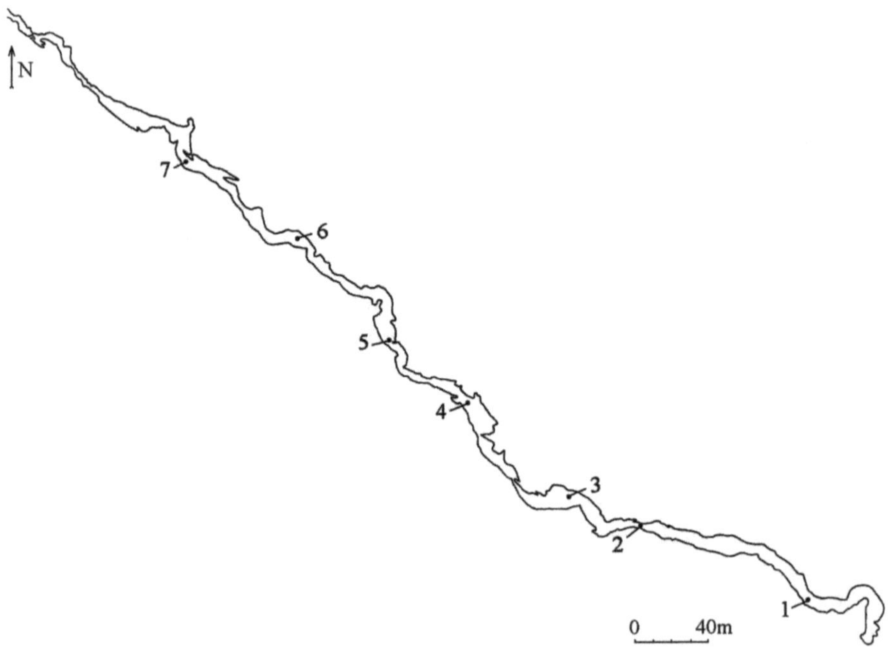

Fig. 3.45 Map of the cave. The numbers are places of observations of deformed speleothems (Kostov 2008)

Fig. 3.46 Inclined massive stalagmites with new vertical speleothems in Lepenitsa Cave

Fig. 3.47 Fractured and inclined massive stalagmite

the strongest known earthquakes until now. These preferred directions are different from the direction of the palaeoflow of the cave sediments—an indication that the fall of the stalagmites was due to the earthquakes. An eventual dating of the sismothems in the future can estimate the age of the seismic event (or events) that caused the deformations in the Lepenitsa Cave.

The epicenters of the earthquakes in the region are connected mainly to the faults of Hremshtitsa fault zone, presented by a group of steep slope faults with general direction 140°. During the early Quaternary in them, as well as the faults of Dospat and Babiak-Grashovska fault zones are performed movements that form in a regional aspect the West Rhodopes block morphostructure (Katskov and Marinova 1992). The absolute dating of deformed speleothems in the cave will probably delineate two or more phases of intense seismotectonic activity associated with movements in the Pleistocene of the described faults.

3.3.2.2 Yamata Cave

The Yamata Cave is located in the area of the Dobrostan Plateau, 38 km S–SE from Plovdiv, Western Rhodopes. The entrance is south facing and is situated at an altitude of 860 m above the deep valley of Mostovska Sushitsa River. The access is relatively difficult, as the only suitable starting point is "Martsiganitsa" hut (1,387 m. a.s.l.) from which the cave is reached with overcome the denivelation of about 550 m (Fig. 3.49).

The Yamata Cave is formed of fine gray-white marble of Dobrostan Marble Formation. Structurally,

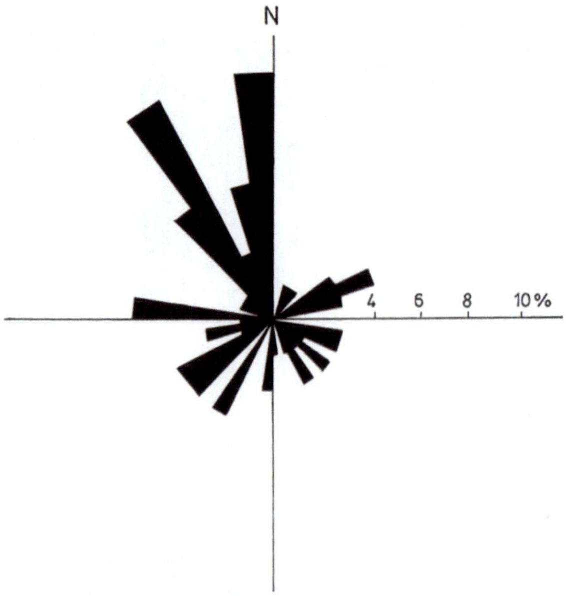

Fig. 3.48 Rose-diagram of the directions of 68 fallen and calcified to the floor stalagmites

the area belongs to North Rhodopean anticline. The sub-parallel faults of the Zaburdo fault zone, intersected by fractures of the Laki fault zone (Kozhuharov et al. 1994) form a characteristic mosaic block and are a precondition to the high degree of marble fracturing and intensive karstification (Fig. 3.50).

In the Dobrostan karst massif have been studied and mapped over 200 caves and pot-holes (Pandev 1993). Definitely dominating the pot-holes, of which the deeper is Druzhba Cave (absolute pit of 130 m., the biggest in Bulgaria), Ivanova Voda Cave (−113 m with 45 m entrance pit), and Zmiin Borun Cave (−108 m). The considerable depths and the wide distribution of vertical cave passages in the area are an indicator for active geodynamic environment during the Quaternary (White 1988).

The cave is dry and consists of one gallery with narrow passages of 20, 40, and 70 m, separating a few chambers. The total length of the gallery is 183 m. The deposits are represented by detritus material

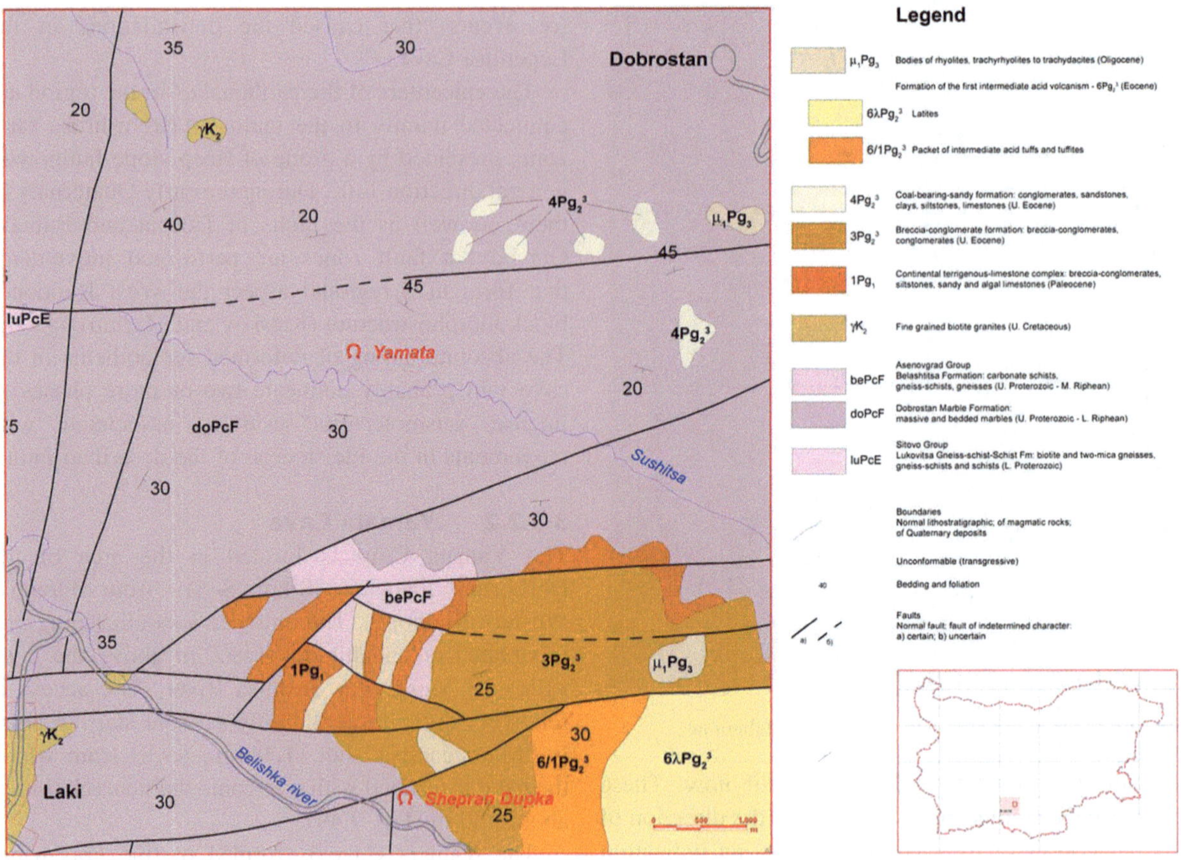

Fig. 3.49 Locations of the Yamata Cave and Shepran Cave and geological map of the area based on sheet "Chepelare" of the Geological Map of Bulgaria in scale 1:100 000 (after Kozhuharov et al. 1994)

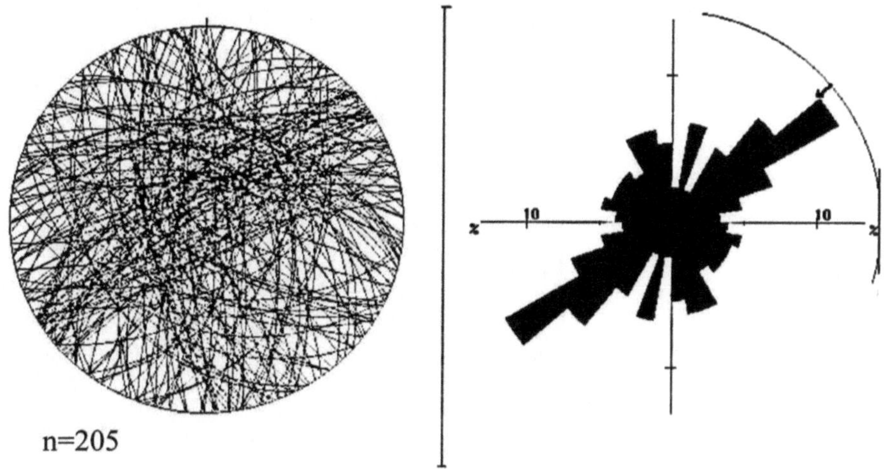

Fig. 3.50 Joint directions in the Dobrostan karst plateau (205 measurements)

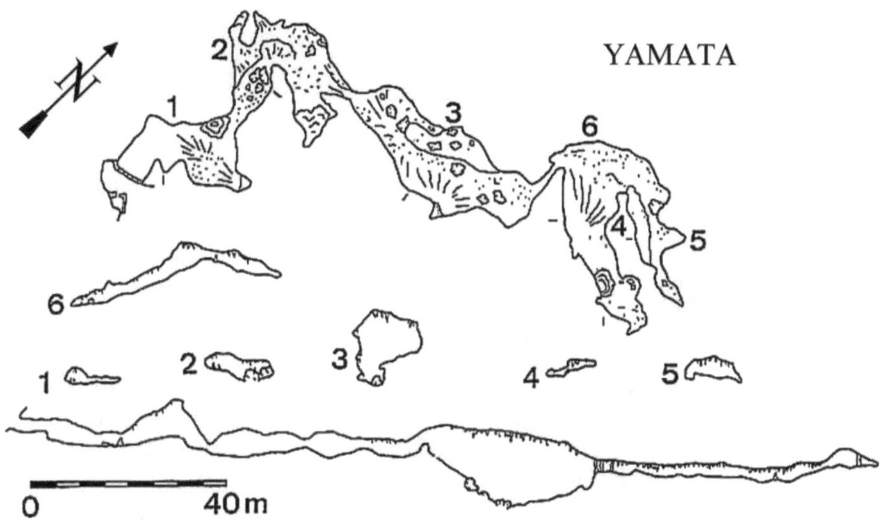

Fig. 3.51 Map, cross sections, and longitudinal profile of Yamata Cave (modified after Pandev 1993)

(breakdown), clay sediments with varied thickness and speleothems, including rare helictites. Yamata Cave was mapped during the National speleoexpedition "Dobrostan 1975" by Boris Kolev from Caving Club "Aida" (Haskovo) and Ivan Dermendzhiev from Caving Club "Galata" (Varna) (Fig. 3.51).

The studies conducted in the cave in September 2005 and May 2007 established multiple deformations of speleothems (Kostov and Shanov 2006; Kostov 2008; Kostov et al. 2009). The deformed formations are widespread and evenly distributed along the cave gallery. The deformations can be separated in the following morphological types:

(A) *Fallen stalagmites*. Their large number compared to the total area of the cave shows the impact of catastrophic tectonic events. The average length of the fallen fragments is 50 cm, but are distributed and massive fallen stalagmites with a height of 1.4 m. All the fallen speleothems are covered with calcite depostis of different thickness.

(B) *Inclined to the north massive stalagmite* with height of 1.7 m and diameter of 1 m with new vertical stalagmite formed on the inclined surface (Fig. 3.52).

(C) *Diagonal crack in massive stalactone* with diameter 1.4 m.

Based on the data on the orientation of the fallen stalagmites was built a rose-diagram (Fig. 3.53). The analysis of the preferred directions indicates three distinct maxima. The three peaks suggest the impact of more than one phase of intense seismotectonic activity during the Quaternary. The discussion on the

Fig. 3.52 K. Kostov is sampling the base of a 26 cm high stalagmite deposited on inclined one (*photo* V. Nikolov)

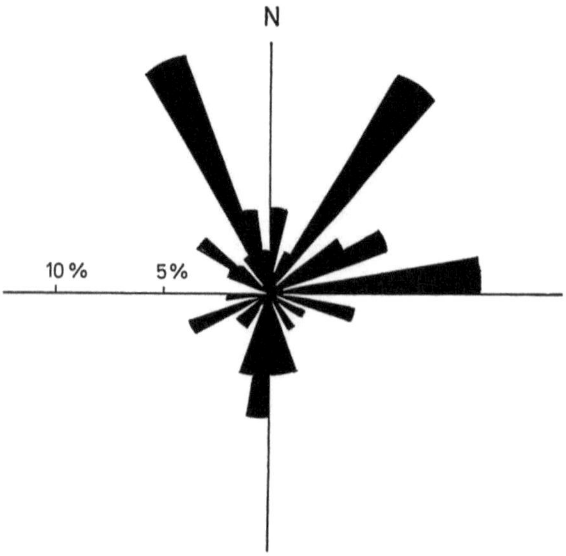

Fig. 3.53 Rose-diagram of the directions of 51 deformed speleothems in Yamata Cave

results will be carried out in comparison with the data from the Shepran Cave located nearby.

In 2007 samples from deformed speleothems were taken for U-series dating. The dating samples were obtained using an electric drill with a steel core barrel COMET M42 (Coronas Bimetal, Spain). The diameter of the taken cores was 12 mm of weight 4–5 g.

The extracted samples were named as follows:

- Yamata1: Core from the base of a small vertical stalagmite with height of 10 cm, deposited on the fallen massive stalagmite 1.4 m long;
- Yamata2: Core from the top of the fallen massive stalagmite with height of 1.4 m, in order to determine the time interval in which the seismotectonic event occured;
- Yamata3: Core from the base of 26 cm high vertical stalagmite deposited on massive inclined stalagmite with height of 1.7 m (Fig. 3.51);
- Yamata4. Core of thick calcite crust covering the top of the flattened "hanging" stalagmite—paleoseismic phenomenon.

Samples were sent to Dr. Gergely Surányi from the radiometric laboratory at the ELTE University in Budapest. The dating was performed by the method of uranium series plasma mass spectrometry (ICP) using plasma ICP-MS spectrometer (Thermo Corp., Germany) at the Institute for isotope studies of the Hungarian Academy of Sciences.

The obtained ages are :

- Yamata1: 48,200 years (lower limit—49,300, upper limit—47 100 years);
- Yamata2: >300,000 years (the large amount of clay components in the sample do not permit accurate analysis and calibration of age);

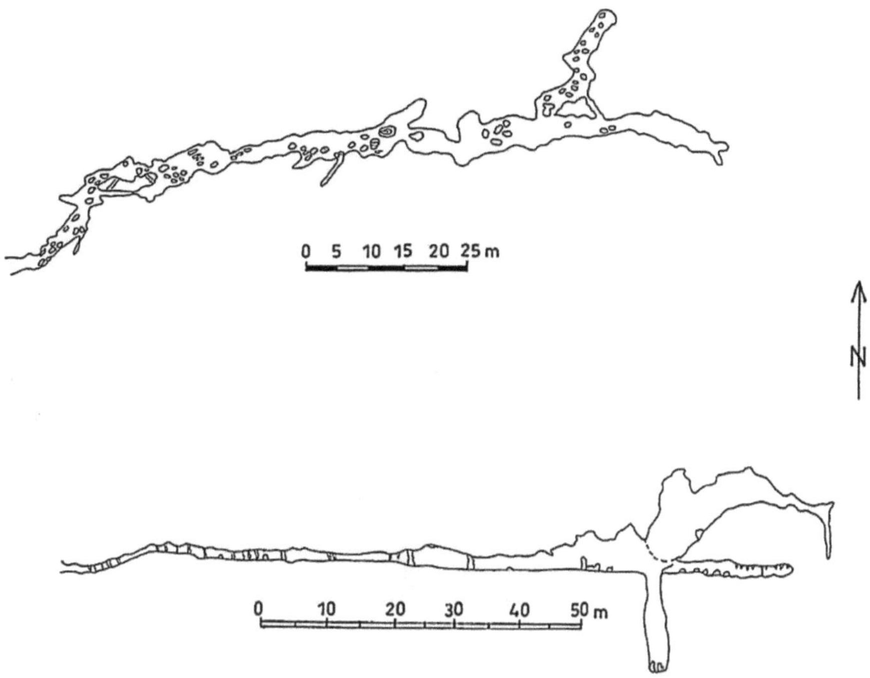

Fig. 3.54 Map and longitudinal profile of Shepran Cave (based on data from Bulgarian Federation of Speleology)

- Yamata3: 229,000 years (lower limit—248,000, upper limit—212 000 years);
- Yamata4: 317,000 years (lower limit—348,000, upper limit—291 000 years).

3.3.2.3 Shepran Cave

The Shepran Cave is located on the southern slopes of the Dobrostan massif, Western Rhodopes, in the valley of Belitsa River, about 1 km northwest of the town of Laki. The cave entrance is at 840 m a.s.l. The cave was surveyed and studied by Chepelare Caving Club "Studenets" in 1963.

Like the other West Rhodopean caves, this cave is formed of Proterozoic marble in Dobrostan Formation (Fig. 3.49).

The Shepran Cave is a dry, fossil cave with interesting morphology, consisting of a gallery of length 260 m and 18 m deep pit at the end (Fig. 3.54). The total denivelation of the cave is 39 m and the volume—3,010 m^3.

During field studies on the cave in the spring of 2001 a large amount of deformed speleothems were found. Prevailing fallen stalagmites and many of the observed forms are covered with thick calcite deposits and younger formations (Fig. 3.55).

The directions of 46 deformed formations are measured. From the constructed rose-diagram is established the existence of three distinct maxima (Fig. 3.56), suggesting the impact of three different phases of tectonic activity.

The orientation in space of the fallen stalagmites in the cave Shepran Cave almost completely coincides with that of the deformation in the Yamata Cave. In both caves are found three well-defined peaks in the preferred direction of the deformed speleothems. The caves are situated in parallel blocks marked by sub-parallel faults of the Zabardo fault zone and are formed in the downthrown blocks of the faults.

These patterns may indicate the conclusion that the deformations in caves located in the downthrown block of a fault are much more significant. This pattern is confirmed by studies in other karst regions in Bulgaria.

The well morphologically expressed sub-parallel Dobrostan fault is marked by 300 m high vertical rock slopes north of Yamata Cave and Shepran Cave. The large number of deformations in these caves can be explained by activation of this fault structure in the Pleistocene with the collapse of the downthrown block where both the caves are situated. On the other hand, in the Ahmetyova Dupka Cave (rich of speleothems), located close to the fault but in its upthrown block, significant amount of deformations are not observed.

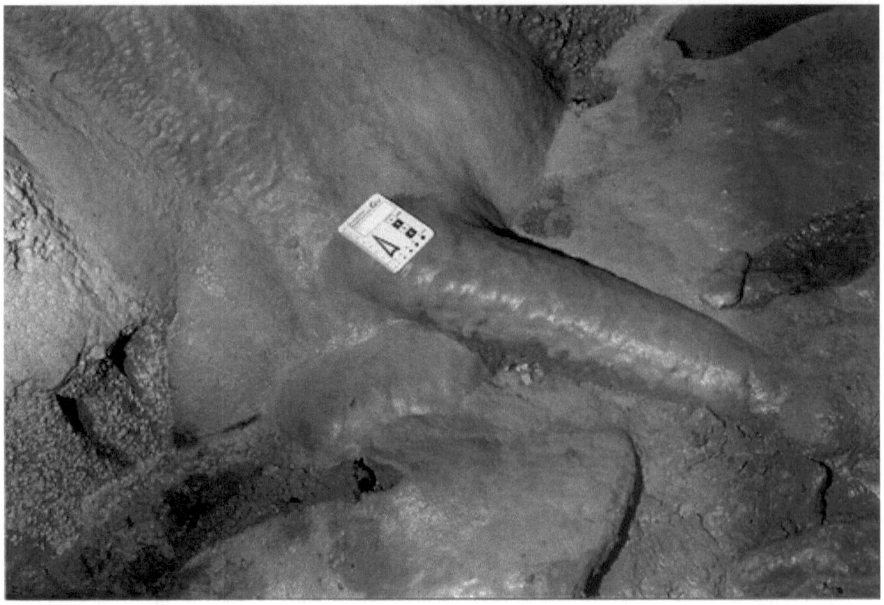

Fig. 3.55 Fallen speleothem covered with thick flowstone in Shepran Cave

The data from the performed dating of deformed stalagmites in Yamata Cave are evidence of the activation of the Dobrostan fault about 300,000 years BP accompanied by a catastrophic effect on the sediments in the cave. The same phase of the seismotectonic activity can explained the deformations in Shepran Cave as well. The active geodynamic behavior of the fault during the Quaternary is confirmed by geomorphological studies in the past (Vaptsarov et al. 1986).

The paleoseimological trench studies performed by the Department of Seismotectonics of the Geological Institute of Bulgarian Academy of Sciences in the last decade on the Popovitsa fault near Plovdiv that generated the catastrophic Popovitsa earthquake ($M = 7.1$, 18.04.1928), found traces of three paleoearthquakes with similar magnitudes to the 1928 event (Yaneva et al. 2004). The excavated trenches are 31–35 km NE from the studied caves.

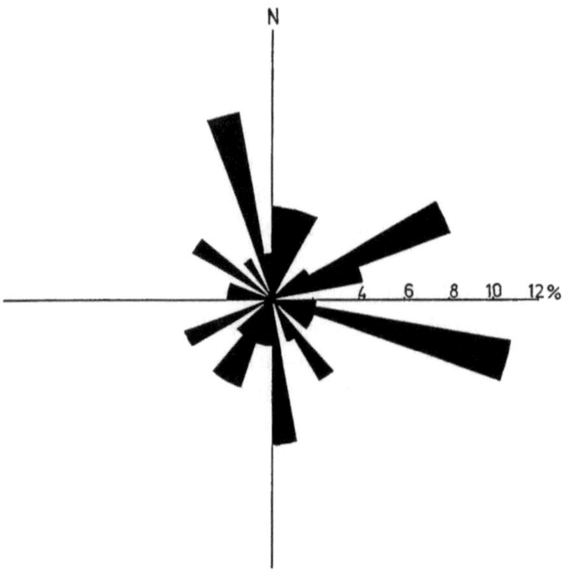

Fig. 3.56 Rose-diagram of the directions of 46 fallen and deformed speleothems in Shepran Cave

3.3.2.4 Snezhanka Cave

Snezhanka is a show cave located on the left slope of the Novomahlenska River, about 5 km southwest of the town of Peshtera in Plovdiv District, Western Rhodopes Mountains.

The cave is developed in highly karstified Proterozoic marble from the Dobrostan Marble Formation and consists of one gallery formed along two basic systems of fractures directed southeast/northwest and south/north. From a morphological point of view, the cave can be divided into six chambers. The studied stalagmites are situated in the biggest one—the Great Hall ($60 \times 40 \times 12$ m).

The length of Snezanka Cave is 230 m, the denivelation is 18 m, and the total surface area of the cave is 3,150 m^2 (Fig. 3.57) The deposits are presented by detritus (breakdown with a maximum block size of 5.5×2.8 m), thin clay sediments, and a variety of different speleothems. The morphological

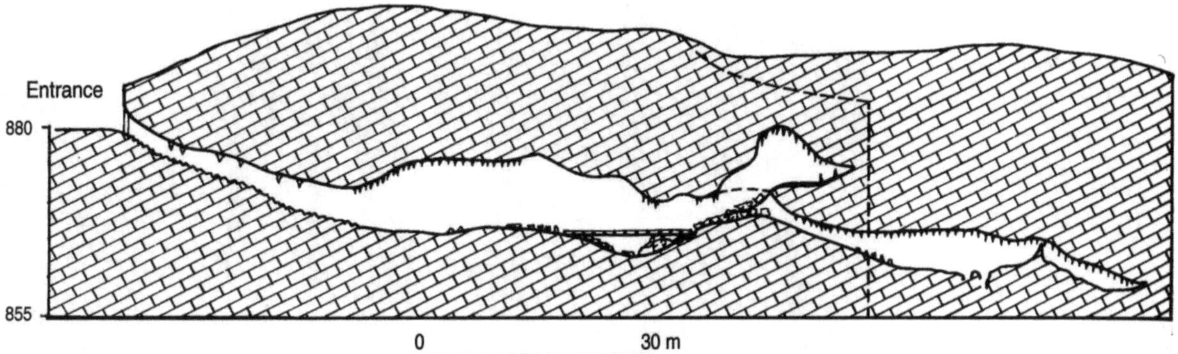

Fig. 3.57 Longitudinal profile of Snezhanka Cave (after Beron et al. 2006)

landmarks of Snezhanka are the highly decorated stalactite ceiling (more than 800,000 stalactites of various shapes) and the availability of thin cylindrical stalagmites suitable for research.

The cave was discovered on January 2, 1961 by cavers G. Kotzev, G. Zlatarev, and B. Evtimov from the town of Peshtera. After archeological excavations and bore-holing in the same year by the team of N. Dzhambazov, artifacts from the early Iron Age and Roman-Thracian era including pottery, the remains of a fireplace and bronze needles were found in the Great Hall of the cave.

Snezhanka Cave has been a show cave for tourists since 1968 and is managed by the Kupena Tourist Society in the town of Peshtera. It was declared as a protected natural monument by Decree 504/ 11.07.1979. Since 1983 the cave has been included in the buffer zone of the Kupena Natural Reserve.

Two thin poplar-like stalagmites – 1 (Fig. 3.58) and 2 (Fig. 3.59)—were investigated (Paskaleva et al. 2008; Gribovszki et al. 2009). Both are 140 cm high but 1 has a complicated shape with a flaring middle part and a very low natural frequency. In both stalagmites the height (H) versus diameter (D) ratios are below ($H/D <$) 20. The diameter of stalagmite 1 ranges between 8.0 and 12.4 cm where the widest is its middle part. The H versus D ratio is between $11.2 < H/D < 17.5$. The results of dimensions measuring are presented in Table 3.3.

The natural frequency measured in Snezhanka stalagmite 1 was exceptionally low, which can be explained by its unusual shape.

According to mechanical test performed on four stalagmite pieces from Snezhanka Cave, the average density was 2.32 g/cm^3, and the standard deviation of eight values measured was 0.15 g/cm^3. Young's modulus was determined using data from a uniaxial compressive strength test. The value was 6,240 MPa; the standard deviation could not be assessed since only one sample was tested. The mean tensile failure stress of the four samples was 2.61 MPa with standard deviation of 0.28 MPa. The Brazilian test was used.

Fig. 3.58 Measurement of Stalagmite 1 in Snezhanka Cave (photo G. Suranyi)

Fig. 3.59 Stalagmite 2 in Snezhanka Cave with attached geophones (*photo* G. Suranyi)

Table 3.3 The parameters of the investigated stalagmites in Snezhanka Cave (Paskaleva et al. 2008)

	H (cm)	^$D_{Average}$ (D_{Min} − D_{Max}) (cm)	Measured natural frequency f_0(Hz)	Theoretical natural frequency f_0(Hz)	Elast modul, E (GPa)	Density (g/cm^3)	Tensile failure stress (MPa)	a_g (m/s^2)
1	2	3	4	5	6	7	8	9
		8.6–12.4						9.2
Snezhanka 1	140	complicated form, middle part flaring	6.1	5.1	6.240	2.32 + 0.1 5	2.61 + 0.28	10.0–14.0
Snezhanka 2	140	11.3 8.3–13.7	19.9	13.2	6.240	2.32 + 0.15	2.61 ± 0.28	10.0–19.0

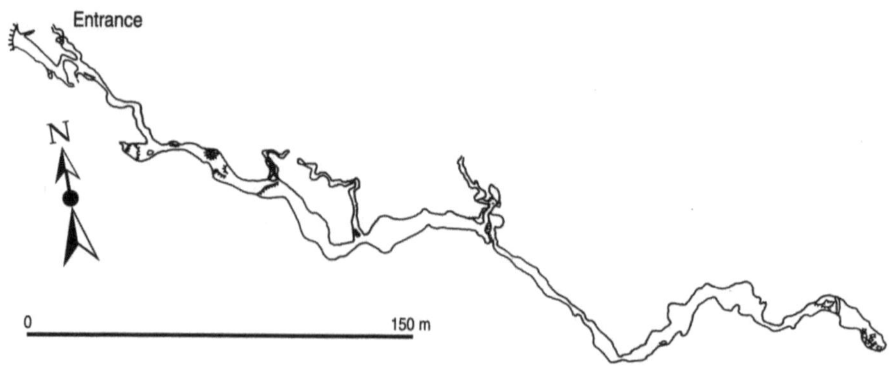

Fig. 3.60 Map of Eminova Cave (after Beron et al. 2006)

Fig. 3.61 Dr. Katalin Gribovszki measuring the natural frequency of the stalagmite in Eminova Cave (*photo* G. Suranyi)

As a result of the field measurements in Snezhanka Cave and the theoretical calculations, the investigated stalagmites were not excited by horizontal acceleration higher than 9.2 m/s^2 in the last few thousand years (Gribovszki et al. 2009).

3.3.2.5 Eminova Cave

Eminova Cave is located in the southern part of the Western Rhodopes Mountains about 5 km northeast of the village of Borino, Smolyan District. The two entrances are situated on the left bank of the deep Kastrakli Gorge and are difficult to access. It is easier to enter the cave from the upper, small, vertical entrance that is on the edge of a 20 m cliff. After penetration through a 25 m narrow passage, one can reach the main gallery of the cave where the stalagmites to be studied are located.

The cave is developed by Proterozoic marbles of the Dobrostan Marble Formation. The total length is 635 m with a denivelation of 35 m. The cave consists of one large northwest/southeast gallery with height of up to 12 m and width of 8–10 m in some places (Fig. 3.60). Typical formations are rich speleothems of different types: stalactites, stalagmites, draperies, flowstones, and sinter lakes (gurs).

Eminova Cave was discovered by Sofia cavers S. Petkov and Ts. Ostromski in 1992. After digging the 25 m narrow passage near the entrance, in the same year it was explored and surveyed by Ts. Ostromski, S. Petkov, M. Zlatkova, Z. Iliev, G. Raichev, and others. The cave is rarely visited and is well protected from negative human impact.

During a visit by the Hungaro-Bulgarian team in 2007 (Dr. K. Gribovszki, T. Czifra, Dr. G. Surányi, Dr. K. Kostov, and G. Nikolov) it was found a suitable thin stalagmite with 117 cm height and 5.1 cm maximal diameter for in situ measurement (Fig. 3.61). The results are presented in Table 3.4.

The peak ground acceleration that can break the studied stalagmite is in the range of 6.0–10.5 m/s^2. Unfortunately, the very thin stalagmites in Eminova and Snezhanka caves are not suitable for sampling for U-series dating. On the base of conducted research, we can suppose that the measured speleothems have not been exited by high horizontal acceleration in the last millennium.

Table 3.4 The parameters of the studied stalagmites in Eminova Cave (Paskaleva et al. 2008)

	H (cm)	$D_{Average}$ ($D_{Min} - D_{Max}$) (cm)	Measured natural frequency f_0(Hz)	Theoretical natural frequency fo(Hz)	Elastic modul, E (GPa)	Density (g/cm^3)	Tensile failure stress (MPa)	a_g (m/s^2)
1	2	3	4	5	6	7	8	9
Eminova Cave	117	4.2 3.0–5.1	17.5	5.7	3.735	2.14 ± 0.25	3.83 ± 0.81	6.0–10.5

References

Angelova D (2003) Karst Types in Bulgaria. Acta Karsologica 32(1):9–18

Angelova D, Benderev A (2000) Karst and tectonics in the Iskar River gorge (Bulgaria). In: Proceedings 4-th symposium on Karst protection. Beograd, 2003, pp 117–122

Angelova D, Benderev A, Baltakov G, Ilieva I, Nenov T (1995) On the karst evolution of the Stara Planina Iskar Gorge. Rev Bulg Geol Soc 56(5):111–124 (In Bulgarian with English abstract)

Angelova D, Benderev A, Kostov K (1999) On the age of the caves in the Stara Planina Iskar gorge, NW Bulgaria. In: Proceedings of the Karst 99 European conference, 10–15 Nov 1999, Grand Causses—Verkors, France, pp 29–35

Aydan Ö (2008) Investigation of the seismic damage caused to the Gunung Sitoli (Tögi-Ndrawa) cave by the 2005 Great Nias earthquake. J Earth Sci Appl Res Centre Hacettepe Univ Turkey 29(1):1–15

Becker HK (1929) Hohle und Erdbeben. Mitt Uber Hohlen und Karstf 1–4:130–133

Becker A, Davenport CA, Eichenberger U, Gilli E, Jeannin P-Y, Lacave C (2006) Speleoseismology: a critical perspective. J Seismol 10:371–388

Benderev A (1989) Karst and karst waters in Ponor Mountain. PhD thesis, NIPI, p 147

Benderev A, Spassov V, Mihaylova B (2006) Impact of hydraulic works on karst grounwaters (Exapmles from Bulgaria). In: BALWOIS 2006, conference on water observation and information system for decision support, 23–26 May 2006, Ohrid, Republic of Macedonia (on CD)

Beron P, Daaliev T, Jalov A (2006) Caves and speleology in Bulgaria. Sofia, Pensoft, p 508

Bini A, Quinif Y, Sules O, Uggeri A (1992) Les mouvements tectoniques recents dans les grottes du Monte Campo dei Fiori (Lombardie, Italie). Karstologia 19:23–30

Blanc J (1985) Phases d, effondrements aux grottes prehistoriques. Karstologia 6:21–28

Bögli A (1969) Neue Anschauungen uber die Rolle von Schichtfugen und Kluften in der Karsthydrographischen Entwicklungen. Geol Rundschau 58:395–408

Braitenberg C, Romeo G, Taccetti Q, Nagy I (2006) The very-broad-band long-base tiltmeters of Grotta Gigante (Trieste, Italy): secular term tilting and the great Sumatra-Andaman islands earthquake of December 26, 2004. J Geodyn 41(1–3):164–174

Braun Y, Kagan E, Bar-Matthews M, Ayalon A, Agnon A (2010) Dating speleoseismites near the Dead Sea Transform and the Carmel Fault: clues to coupling of a plate boundary and its branch. Israel J Earth Sci 58(3–4):257–273

Briestenský M, Košťák B, Stemberk J, Petro L, Vozár J, Fojtíková L (2010) Active tectonic fault microdisplacement analyses: a comparison of results from surface and underground monitoring in Western Slovakia. Acta Geodyn Geomater 7, 4(160):387–397

Bruxelles L, Guendon J-L, Quinif Y (1998) L'aven de la Portalerie (Larzac, Aveyron). Enregistrement d'évènements sismo-tectoniques dans les Grands Causses par les concrétions de l'endokarst, Actes du colloque "Géomorphologie quantitative et paléogéomorphologie dans les karsts du domaine méditerranéen", Massif de la Sainte-Baume, 12-13 octobre 1997, In: Etudes de Géographie Physique, travaux 1998, supplément au n°27, 1998, pp 45–56

Cadorin J-F, Jongmans D, Plumier A, Quinif Y, Camelbeeck T (2000) Modelling speleothems rupture. In: Proceedings of the "Han 2000" workshop, 13–17 March 2000, Han-sur-Lesse, Belgium, pp 27–30

Cadorin J-F, Jongmans D, Plumier A, Camelbeeck T, Delaby S, Quinif Y (2001) Modelling of speleothem failure in the Hotton Cave (Belgium). Is the failure earthquake induced? Netherlands J Geosci 80(3–4):315–321

Cai WX, Van Ruymbeke M, Tang S, Ducarme B, Du W, Flick J (1989) Cooperative projects in geodynamical instrumentation between the State Seismological Bureau (China) and the Royal Observatory of Belgium. In: Proceedings of the Xth international symposium on Earth Tides, Helsinki, July 31–August 5 1989

Camelbeeck T (1998) Speleothems as palaeoseismic indicators: the point of view of a seismologist. In: Contributions to the international symposium on Karst & Tectonics, 9–12 March 1998, Han-sur-Lesse, Belgium, pp 23–24

Colman S, Perce K, Birkeland P (1987) Suggested terminology for Quaternary dating methods. Quatern Res 28:314–319

Delaby S (1999) Etude statistique de l'enregistrement paleosismique par les speleothemes. L'exemple de la Grotte de Hotton (Belgique). In: Proceedings of the "Karst 99" European Confernce 10–15 Nov 1999, Grand Causses–Vercors, France, pp 73–76

Delaby S (2000) Palaeoseismic investigations in Belgium caves. In: Proceedings of the "Han 2000" workshop, 13–17 Nov 2000, Han-sur-Lesse, Belgium, pp 45–48

Delange P, Guendon J-L (1998) Mise en evidence d'un seisme au Pleistocene dans la grotte de Ribiere (Bouches-du-Rhone). Contributions to the International Symposium on Karst & Tectonics, 9–12 Nov 1998, Han-sur-Lesse, Belgium, 4 p

Dobrev N, Avramova-Tacheva E (1997) Analysis and prognostication of monitored rock deformations. In: Proceedings of the IAEG conference, Athens 1997, Balkema, Rotterdam, pp 613–618

Dublyansky V (1995) Signs of intense earthquakes in karst regions (with special reference to the Mountain Crimea). Geomorphology, Russian Academy of Sciences 1:38–46 (in Russian with English summary)

Dublyansky V, Molodih I (1972) The seismicity of Crimea according to data from karstological and archaeological studies. In: Problems of hydrogeology and groundwater science, Kiev, Naukova Dumka, pp 43–51 (in Russian)

Dubois P, Grellet B (1997) Les concretions des grottes enregistrent climats et seismes. Pour la Science, 231, Janvier 1997:28–35

Forti P (1997) Spelothems and Earthquakes. In: Hill CP, Forti (eds) Cave miinerals of the world, Nat Spel Soc, pp 284–285

Forti P (1998) Seismotectonic and paleoseismic studies from speleothems: the state of the art. In: Contributions to the international symposium on Karst & Tectonics, 9–12 March 1998, Han-sur-Lesse, Belgium, pp 79–81

Forti P, Petrini V, Postpischl D (1983) Ricostruzione di fenomeni paleosismici. da strutture carsiche. Rend Soc Geol It 4:563–569

Forti P, Postpischl D (1981a) Neotectonic data from stalagmites: sampling and analysing technics. In: Proceedings of the European region conference on Speleology, 22–28 Nov 1980, Sofia, vol 2, pp 25–33
Forti P, Postpischl D (1981b) The hypothesis of the induced activity of the faults as a result of a statistical analysis of stalagmites. In: Proceedings of the European region conference on Speleology, 22–28 Nov 1980, Sofia, vol 2, pp 34–39
Forti P, Postpischl D (1984) Seismotectonics and paleoseismic analyses using karst sediments. Mar Geol 55:145–161
Forti P, Postpischl D (1985) Relations entre tremblements de terre et deviation des axes des stalagmites. Sottoterra 72:21–27
Forti P, Postpischl D (1986) May the growth axes of stalagmites be considered as recorders of historic and prehistoric earthquakes?—Preliminary results from the Bologna karst area (Italy). In: Proceedings on engineering geology problems in Seismic area, Bari, 1, pp 183–193
Forti P, Postpischl D (1987) Seismotectonics and radiometric dating of karst sediments. In: Proceedings of the historical seismity of central-eastern mediterranian region, 27–29 Dec 1987, Rome, pp 321–332
Frangov G, Ivanov P, Dobrev N, Iliev I (1992) Stability problems of the rock monument "Madara horseman". In: Proceedings of the 7th international congress on deterioration and conservation of stone, Lisboa, Portugal, pp 1425–1435
Garasic M (1981) Neotectonics in some of the speleological objects in Yugoslavia. Proceedings of the 8th International Congress on Speleology, Bowling Green, USA, vol 1 and 2, pp 148–150
Garasic M, Cvijanovic D (1985) Speleological phenomena and seismic activity in Dinaric karst area in Yugoslavia. In: Comm. 9th Congress on International Espel. Barselona, pp 94–95
Gilli E (1986) Neotectonique dans les massifs karstiques, un exemple dans les Prealpes de Nice: la grotte des Deux Gourdes. Karstologia 8:50–52
Gilli E (1995) Recording of earth movements in karst. In: 5th international conference on Seizm. Zonation, 17–19 Dec 1995, Nice, Ouest Edit., Nantes, pp 1305–1314
Gilli E (1996) Effets des seismes dans l'endokarst. Application aux etudes de sismicite historique. In: Geomorphologie, risques naturels et amenagement. Lab Rev An Spat Raoul Blanchard 38–39, Nice, pp 121–132
Gilli E (1998) Approche de la paleosismicite de Monaco par l'etude d'un plancher stalgmitiqie de la Grotte de l'Observatoire. In: Contributions to the international symposium on Karst & Tectonics, 9–12 March 1998, Han-sur-Lesse, Belgium, pp 83–85
Gilli E (1999) Rupture de speleothemes par fluage d, un remplissage endokarstique. L, exemple de la grotte de Ribiere (Bouches-du-Rhone). C. R. Acad. Sci Paris, Sciences de la terre et des planetes 329:807–813
Gilli E (2005) Review on the use of natural cave speleothems as palaeoseismic or neotectonic indicators. Comput Rendus Geosci 337:1208–1215
Gilli E, Levret A, Sollogoub P, Delange P (1998) Recherches sur le seisme du 18 fevrier 1996 dans les grottes de la region de Saint-Paul-de-Fenouillet (Pyrenees-Orientales, France). In: Contributions to the international symposium on Karst & Tectonics, 9–12 March 1998, Han-sur-Lesse, Belgium, pp 99–102
Gilli E, Serface R (1999) Evidence of palaeoseismicity in the caves of Arizona and New Mexico (USA). C. R. Acad Sci Paris, Sciences de la terre et des planetes 329:31–37
Gosar A., Sebela S, Kostak B, Stemberk J (2011) On the state of the TM 71 extensometer monitoring in Slovenia: Seven years of micro-tectonic displacement measurements. Acta Geodyn Geomater 8, 4(164):389–402
Gospodarich R (1968) Podrti kapnikiv Postejtiski Jami. Nase Jame 9(1967):15–31
Gospodarich R (1977) Collapsing of spelothems in Postojna Cave system. In: Proceedings on 7th International Speleology Congress, Sheffield, UK, pp 223–240
Gospodarich R (1981) Sinter generations in classical karst of Slovenia. Acta Carsologica 9:90–110
Gribovszki K, Bokelmann G, Szeidovitz G, Varga P, Paskaleva I, Brimich L, Kovács K, Mónus P (2013a) Comprehensive investigation of intact, vulnerable stalagmites to estimate an upper limit on prehistoric horizontal ground acceleration. In: Adam C, Heuer R, Lenhardt W, Schranz C (eds) Vienna congress on recent advances in earthquake engineering and structural dynamics 2013 (VEESD 2013), 28–30 Aug 2013, Vienna, Austria, Paper No 445
Gribovszki K, Brimich L, Varga P, Kovács K, Shen C-C, Kele S, Török Á, Novák A (2013b) Estimation of an upper limit on prehistoric horizontal peak ground acceleration using the parameters of intact stalagmites and the mechanical properties of broken stalagmites in Domica Cave, Slovakia. In: Geophysical Research Abstracts, vol 15, EGU2013-11541, 2013 EGU General Assembly 2013
Gribovszki K, Kovács K, Mónus P, Chuan-Chou Shen, Török Á, Brimich L, Varga P, Novák A, Kele S (2013c) Estimation of an upper limit on prehistoric peak ground acceleration using the parameters of intact stalagmites and the mechanical properties of broken stalagmites in Domica cave, Slovakia. In: 9th scientific conference "Research, Use and Protection of Caves", Liptovská Sielnica, 23–26 Nov 2013, pp 48–49
Gribovszki K, Paskaleva I, Kostov K, Varga P, Nikolov G (2009) Estimating an upper limit on prehistoric peak ground acceleration using the parameters of intact speleothems in caves in southwestern Bulgaria. In: Zaicenco A, Craifaleanu I, Paskaleva I (eds) Harmonization of seismic hazard in Vrancea zone with special emphasis on seismic risk reduction, Springer, pp. 287-308
Grigorova E, Grigorov B (1964) The epicenters and the seismic lines in Bulgaria. BAS Publishing house, Sofia, p 84 (in Bulgarian with English summary)
Hohenwart F Graf von (1832) Wegweiser für die Wanderer in der berühmten Adelsberger und Kronprinz Ferdinands-Grotte bey Adelsberg in Krain; als Erklärung der von Herrn Aloys Schaffenrath, k.k. Kreis-Ingenieur in Adelsberg, gezeichneten Ansichten dieser Grotte, Wien, JP Sollinger, 3 Hefte, 16, 9, 14 S. und 21 Tafeln (Nachdruck 1978: Šajn S, Hrsg, Habe F, Šlenc J, Vorwort: Postojnska Jama; 1. Aufl . Postojnska Jama THO; Postojna; 32 S. Einleitung; Querfolio)
Hrischev H (1966) Litostratygraphy of the Lovech Urgonian Group. News Geol Inst 15:231–246

Hrischev H, Nedialkova L (1992) Explanatory note to the sheet "Sevlievo" of the Geological map of Bulgaria in scale 1:100 000. BAS Publishing house, Sofia, p 56

Ilieva I, Zdravkov I, Benderev A (1981) The role of tectonic movements for karst development around Cherepish railway station. In: Proceedings of Speleology conference, 4–6 May 1979, Rousse, Bulgaria, pp 209–214 (In Bulgarian)

Jeannin P-Y (1990) Neotectonique dans le karst du nord du Lac de Thoune (Suisse). Eclogae Geol Helvet 82(2):323–342

Jordanov M, Popov N, Mandov G, Nedialkova S (1961) On the tectonics of the Vratsa Range. Rev Bulg Geol Soc 22(2):205–216 (in Bulgarian with German abstract)

Kagan EJ (2002a) Dating large earthquakes using damaged cave deposits, Soreq and Har-Tuv caves, Central Israel. Geological Survey of Israel report GSI/19/02

Kagan EJ (2002b) Dating large earthquakes using damaged cave deposits, Soreq and Har-Tuv caves, Central Israel. M.Sc. thesis. Hebrew University of Jerusalem

Kagan EJ, Agnon A, Bar-Matthews M, Ayalon A (2002a) Cave deposits as recorders of paleoseismicity: a record from two caves located 60 km West of the Dead Sea transform (Jerusalem, Israel). In: Leroy S, Stewart IS (eds) Environmental catastrophes and recovery in the holocene. Abstracts Volume. Brunel University, west London (UK), 28 Aug–2 Sept 2002

Kagan EJ, Bar-Matthews M, Ayalon A, Agnon A (2002b) A wiggle-matching technique applied to the dating of damaged cave deposits and compilation of a long- term paleoseismic record, Soreq Cave, Israel. In: Goldschmidt conference abstracts. Davos, Switzerland, 18–23 Aug 2002

Kagan EJ, Agnon A, Bar-Matthews M, Ayalon A (2005) Dating large infrequent earthquakes by damaged cave deposits. Geology 33(4):261–264

Kagan EJ, Bar-Matthews M, Ayalon A, Agnon A (2006) Damaged speleothems record 200 000 years of earthquake activity: Dead Sea transform region. Arch Clim Change Karst Waters Inst Spec Publ 10:33–34

Katskov N, Marinova R (1992) Explanatory note to the sheet "Belitsa" of the geological map of Bulgaria in scale 1: 100 000. BAS Publishing house, Sofia, p 43

Kempe S (2004) Natural speleothem damage in Postojnska Jama (Slovenia), caused by glacial cave ice? A first assessment. Acta Carsologica 33(1):265–289

Kis M, Detzky G, Koppán A (2012) 3D Finite element modelling for the investigation of the cavity effect in extensometric rock-deformation measurements. In: Gephysical Research Abstracts, vol 14, EGU2012-9671. http://meetingorganizer.copernicus.org/EGU2012/EGU2012-9671.pdf

Klimeš J, Rowberry MD, Blahůt J, Košťák B, Rybář J, Hartvich F, Briestenský M, Stemberk J, Štěpančíková P (2012) The monitoring of slow moving landslides and assessment of stabilisation measures using an optical-mechanical crack gauge. Landslides 9:407–415

Košťák B (1969) A new device for in situ movement detection and measurement. Exp Mech 9:374–379

Košťák B (1991) Combined indicator using Moiré technique. In: Proceedings on 3rd international symposium in field measurements in geomechanics, 9–11 Sept 1991, Oslo, pp 53–60

Kostak B, Dobrev N, Zika P, Ivanov P (1998) Joint monitoring on a rock face bearing an historical bas-relief. Q J Eng Geol Lond 30(1):37–46

Kostov K (1997) Karst morphology in Bazovski part of Vratsa Mountain (West Stara Planina, NW Bulgaria). In: Proceedings on 12 international congress on speleology, 10–17 Aug 1997, La Chaux-de-Fonds, Switzerland, 1, pp 409–412

Kostov K (1999) Data about Quaternary tectonic activity in Labirinta Cave in Stara Planina Iskar Gorge, NW Bulgaria. Theor Appl Karstol 11–12:17–22

Kostov K (2000) Sights of intense earthquakes in Bulgarian mezokarstic forms. In: Abstracts on 31st international geological congress, Rio de Janeiro, Brazil (on CD)

Kostov K (2001) Palaeoseismological studies using speleothems in Bulgarian caves. In: Fifth international conference on geomorphology, Tokyo, Japan, Abstracts of conference papers, transactions Japanese geomorphological union, vol 22, 4, p 131

Kostov K (2002) Speleothems as paleoseismic indicators: examples from Bulgaria. In: Leroy S, Stewart IS (eds) Environmental catastrophes and recoveries in the holocene conference, 29 Aug–2 Sept 2002. Brunel University, London, pp 43–44

Kostov K (2004) Paleoseismological investigations in karstic caves: examples from Bulgaria. In: Abstracts 32nd international geological congress, Florence (on CD)

Kostov K (2008) Traces of paleoseismicity in karst terrains. Ph.D. Thesis, Geological Institute—BAS, Sofia, p 191

Kostov K, Shanov S (2006) Records of catastrophic events on speleothems: case studies from the Rhodopes, Southern Bulgaria. In: Karst Waters Institute Special Publication 10, USA (Proceedings of the IVth climate change: the Karst record symposium, 26–29 May 2006, Baile Herculane, Romania), pp 36–38

Kostov K, Shanov S, Kourtev K, Nikolov G, Boykova A, Benderev A (2000) Broken speleothems as indicators of palaeoseismic activity: an example from Lepenitsa cave in the Western Rhodope Mts., South Bulgaria, In: Abstracts on geological conference, 11–13 Oct 2000, Sofia, pp 217–218

Kostov K, Shanov S, Surányi G (2009) Palaeoseismological investigations using speleothems: case study of two caves in Rhodopes Mountains, Southern Bulgaria. In: 1st INQUA IGCP 567 International workshop on earthquake archaeology and palaeoseismology, 07–13 Nov 2009, Baelo Claudia, Spain, pp 76–78

Kozhuharov D, Kozhuharova E, Marinova R, Katskov N (1994) Explanatory note to the sheet "Chepelare" of the Geological Map of Bulgaria in scale 1:100 000. BAS Publishing house, Sofia, p 94

Kozhuharova E, Kozhuharov D (1962) Studies on the rocks and the construction of the North Rhodopean Anticline in Asenovgrad region. News Geological Inst 11:125–162 (in Bulgarian)

Lacave C, Levret A, Koller M (2000) Measurements of natural frequencies and dampling of speleothems. In: Proceedings on 12th World conference earthquake engineering, Auckland, New Zealand, paper 2118

Lacave C, Koller MG, Eichenberger M, Jeannin P-Y (2003) Prevention of speleothem rupture during nearby construction. Environ Geol 43:892–900

Lacave C, Koller MG, Ezogue JJ (2004) What can be concluded about seismic history from broken and unbroken speleothems? J Earthquake Eng 8(3):431–455

Lemeille F, Cushing M, Carbon D, Grellet B, Bitterli T, Flenoc C, Innocent C (1999) Co-seismic ruptures and deformations recorded by speleothems in the epicentral zone of the Basel earthquake. Geodin Acta 12(3–4):179–191

Lundberg J, McFarlane DA (2012) Cryogenic fracturing of calcite flowstone in caves: theoretical considerations and field observations in Kents Cavern, Devon, UK. Int J Speleol 41:307–316

Mircheva M, Iliev Z, Kostov K (2004) The caves on the Madara Plateau. "Iskar", Sofia, 42 p. (in Bulgarian with English summary)

Mocchiutti A (2004) Evidenze morfologiche di movimenti tettonici recenti ed attuali nelle cavita delle Alpi Carniche. Il fenomeno carsico dell Alpi Carniche. Mem Ist Ital Spel S II 15:47–50

Moser M, Geyer M (1979) Seismospelaologie—Erdbebenzerstorungen in Hohlen am Beispiel des Gailoches bei Oberfellendorf (Oberfranken, Bayern). Die Hohle 30(4):89–102

Nikolov T, Monov B, Mitov K, Petkov P (1972) Lithostratigraphy of the Vratsa Urgonian group. Rev Bulg Geol Soc 33(2):205–216

Orozova-Stanishkova I, Costa G, Vaccari F, Suhadolc P (1996) Deterministic estaimates of the seismic hazard in Bulgaria. Spl Publ Geol Soc Greece 6:183–185

Palmer A (1984) Geomorphologic interpretation of karst features. Groundwater as geomorphic agent. Allen and Unwin, Boston, pp 174–209

Pandev R (1993) Dobrostan karst massif—caves and pot-holes. IMN Publishing House, Plovdiv, p 128

Panno SV, Lundstrom C, Hackley KC, Curry BB, Fouke B, Zhang Z (2009) Major earthquakes recorded by speleothems in Midwestern U.S. caves. Bull Seismol Soc Am 99(4):2147–2154

Paskalev M, Benderev A, Shanov S (1992) Tectonic conditions in the area of the Iskrets karst springs. Rev Bulgarian Geol Soc LIII 2:69–81 (in Bulgarian with English abstract)

Paskaleva I, Szeidovitz I, Kostov K, Koleva G, Nikolov G, Gribovszki K, Czifra T (2006) Calculating the peak ground horizontal acceleration generated by paleoearthquakes from failure tensile stress of speleothems. In: Proceedings on international conference on civil engineering, design and construction, 14–16 Sept 2006, Varna, Bulgaria, pp 281–286

Paskaleva I, Gribovszki K, Kostov K, Varga P, Nikolov G (2008) Peak ground acceleration assessment using the parameters of intact speleothems in caves situated in NW and SW Bulgaria. In: Proceedings international conference on civil engineering design and construction, 12–14 September 2008, Varna, Bulgaria, pp 249–264

Pérez-López R, Rodrígues-Pascua MA, Giner-Robles JL, Martínez-Díaz JJ, Maros-Nuez A, Silva PG, Bejar M, Calvo JP (2009) Speleoseismology and palaeoseismicity of Benis Cave (Murcia, SE Spain): coseismic effects of the 1999 Mula earthquake (m_b 4.8). Geol Soc Lond Spec Publ 316:207–216

Petrov P (1983) Hydrogeological events induced by the earthquake (seismohydrogeological phenomena). In: The Vrancea Earthquake—1977. Consequences in Bulgaria. BAS Publishing house, Sofia, pp 96–112 (in Bulgarian)

Pons-Branchu E, Hamelin B, Brulhet J, Bruxelles L (2004) Speleothem rupture in karst: tectonic or climatic origin? U-Th dating of rupture events in Salamandre Cave (Gard, Southeastern France). Bull Soc Geol France 175(5):473–479

Popov V (1979) Saeva Dupka. Sofia, Meditzina i Fizkultura, p 24 (in Bulgarian)

Postpischl D, Agostini S, Forti P (1990) Grotta dei Cervi (Pietrasecca, Abruzzo): Studio dei principali terremoti preistorici dell, area carseolana dall, analisi di strutture carsiche. Rend Soc Geol It 13:57–64

Postpischl D, Agostini S, Forti P, Quinif Y (1991) Paleosismicity from karst sediments: the Grotta del Cervo cave case study (Central Italy). Tectonophysics 193:33–34

Quinif Y (1996) Enregistrement et datation des effets sismotectoniques par l, etude des speleothemes. Ann Soc Geol Belgique 119(1):1–13

Quinif Y (1999) Etude d, un sismotheme dans le Reseau Sud de la Grotte de Han-sur-Lesse. Speleochronos 10:33–46

Quinif Y (2000) Karst and sismotectonics: speleothems as paleoseismic indicators. Proceedings on "Han 2000" workshop, 13–17 March 2000, Han-sur-Lesse, Belgium, pp 121–124

Radev J (1915) Karst forms in Western Stara Planina. Ann Sofia University, Hist Phil Fac 149:10–11 (in Bulgarian)

Sancho C, Pena J-L, Mikkan R, Osacar C, Quinif Y (2003) Morphological and speleothemic development in Brujas Cave (Southern Andean Range, Argentine): palaeoenvironmental significance. Geomorphology 57(3–4):367–384

Schillat B (1965) Aufschluss. Nachweis von Erdbeben in Hohlen 16(1):133–136

Schillat B (1970) Discussion of "the detection of prehistoric earthquakes from fractured cave structures". Caves Karst 12(5):33–36

Šebela S (2008) Broken speleothems as indicators of tectonic movements. Acta Carsologica 37(1):51–62

Šebela S (2009) 3D monitoring of active tectonic movements in Slovene karst caves. Geološki zbornik 20:152–155

Šebela S (2010) Effects of earthquakes in Postojna cave. Acta Carsologica 39(3):597–604

Šebela S, Gosar A, Košťák B, Stemberk J (2005) Active tectonic structures in the W part of Slovenia—Setting of micro-deformation monitoring net. Acta Geodyn Geomater 2(1):45–57

Shanov S, Kourtev K, Kostov K, Nikolov G, Boykova A, Benderev A (1998) Palaeoseismological traces in the Lepenitsa cave, Velingrad district, South Bulgaria. Pro-GEO'98 International conference on Abstracts volume. Belogradchik, Bulgaria, pp 53–54

Shanov S, Kourtev K, Kostov K, Nikolov G, Boykova A, Benderev A (2001) Palaeoseismological traces in the Lepenitsa Cave, Velingrad district, South Bulgaria. RIVIERA 2000—Tectonicue active et geomorphologique. Villefranche-sur-Mer, France, Revue d, Analyse Spatiale—No special, pp 151–154

Skorpil H, Skorpil K (1898) Sources et Pertes des Eaux en Bulgarie. Mémoires de la Société de Spéléologie, III, 15, Paris, p 46

Sowers J, Noller J (1997) Quaternary geochronologic methods. In: Sowers JM, Noller JS, Lettis WR (ed) Dating and earthquakes: review of quaternary geochronology and its application to paleoseismology. William Lettis & Associates, Inc., Walnut Creek, pp 1–9

Spöcker RG (1933) Ursachen und Formen des Höhlenzerfalls im Frankenjura. Speläol Jb 13(14):66–83

Szeidovitz G, Leel-Ossy S, Suranyi G, Czifra T, Gribovszki K (2005) Calculating the peak ground horizontal acceleration generated by paleoearthquakes from failure tensile stress of speleothems. Magyar Geofizika 46(3):91–101

Szeidovitz G, Paskaleva I, Gribovszki K, Kostov K, Surányi G, Varga P, Nikolov G (2008) Estimation of an upper limit on prehistoric peak ground acceleration using the parameters of intact speleothems in caves situated at the western part of Balkan Mountain range, north-west Bulgaria. Acta Geod Geoph Hung 43(2–3):249–266

Tronkov D (1965) Tectonic structure and analysis of the structures of the Vratsa block of Western Stara Planina Mountains. Plastic deformations near the fault planes. Pap Geol Bulgaria Stratigr Tecton 6:217–251 (in Bulgarian)

Tronkov D (1981) The stratigraphy of the Triassic system in parts of the Western Sredna Gora (West Bulgaria). Geologica Balkanica 11:3–20 (in Russian)

Tzankov T, Nikolov G (1996) Recent bilateral listric destruction in West Stara Planina Mountain and the West Forebalkan, Bulgaria. Compt Rend Acad Bulg Sci 49(4):53–54

Urbani F (2002) Espeleotemas rotadas en las cuevas de Guanasna, EstadoMiranda, Venezuela: Estructuras de probable origen paleosismico. Bol Soc Venezolana Espeleol 36:17–20

Vanara N (1997) Datation d, un phenomene catastrophique dans la grotte-tunnel d'Alzaleguy (massif des Arbailles, Pyrenees-Atlantiques, France). Speleochronos 8:13–22

Vanneste K, Verbeeck K, Camelbeeck T, Paulissen E, Meghraoul M, Rennardy F, Jonghmans D, Frechen M (2001) Surface-rupturing history of the Bree fault scarp, Roer Valley graben: evidence for six events since the late Pleistocene. J Seismol 5:329–359

Vaptsarov I, Mishev K, Kochneva N, Aleksiev G (1986) Tectogenic elements in the present-day relief of South Bulgaria and deciphering of superimposed deep structures. Geol Balcanica 16(6):3–19.

White W (1988) Geomorphology and hydrology of karst terrains. Oxford University Press, Oxford, p 288

Wojcik Z, Zvolinski S (1959) Mlode przesunecia tektoniczne w jaskiniach tatrzanskih. Acta Geol Polonica 9(2):318–341

Yaneva M, Radulov A, Nikolov G, Shanov S, Camelbeeck T, Nikolov N, Mitev A, Kostov K, Vaneste K, Verbeeck K (2004) Geological records for paleoseismicity of Popovitsa fault in a trench near Popovitsa, Southern Bulgaria. In: Proceedings on conference of the Bulgarian geological society "Geology 2004", Sofia, 16–17 Dec 2004, pp 107–109

Yovcev Y (1971) Tectonic Structure of Bulgaria. In: Yovcev Y (ed) Technika, Sofia, p 558 (in Bulgarian with English summary)

Zacharov M (1984) Research of the geological-structural circumstances and deformations in Jasovská Cave. Slovenský Kras 22:69–94 (in Slovak with English summary)

Zlatkova M (1987) Methods of applied geophysics for the search of underground karst cavities. In: 5th National conference of speleology, 28–30 Aug 1987, Sofia, pp 24–32 (in Bulgarian)

Glossary

Abyss Sub-vertical cavity of extremely great depth

Active cave A cave containing a running stream

Anisotropy A variance of physical characteristics of mass depending on the direction of their measurement

Anomaly A deviation of the characteristics from the normally expected

Aquifer, karst An aquifer in which the water flow is appreciable trough joints, faults and cavities enlarged by dissolution

Aquifer, confined An aquifer bounded above and below by confining rocks of distinctly lower permeability than that of the aquifer itself

Aquiclude Porous rock formation capable of storing water without sufficient transmitting rates to furnish an appreciable supply for spring

Aquifuge Porous rock formation without interconnected openings or interstices unable neither to store nor to transmit water

Aquitard A confining bed retarding, but not preventing the water flow to or from an adjacent aquifer

Allochthon Rock formation originating from different locality than the one in which it has been deposed

Autochthon Rock formation originating and deposed at about the same location

Barrier A geological formation or structure having become impervious to ground water flow

Base of karstification Level below which karstification has not occured

Basin Hydrogeographic unit receiving water from precipitations and discharging it trough one or more springs

Brittle deformation A property of materials for sudden failure; complete loss of cohesion across one or more planes

Catchment An area or a depression that collects and drains rainwater

Cave A single, different lengths accessible passage or an extensive complex network of underground galleries

Cavern An underground opening in soluble rocks

Cavern system An underground network of passages, chambers, galleries and other cavities with or without underground river(s)

Conjugated joints Two sets of joints that are formed under the same stress conditions (during the same deformational event) *Synonym: Joint system*

Crack A small fracture with respect to the scale of the volume in which it occurs

S. Shanov and K. Kostov, *Dynamic Tectonics and Karst*, Cave and Karst Systems of the World,
DOI: 10.1007/978-3-662-43992-0, © Springer-Verlag Berlin Heidelberg 2015

Dip The angle between an inclined surface (bedding plane, fracture plane) and the horizontal

Dip direction The angle between the geographical north and the sense of dipping of an inclined surface (bedding plane, fracture plane)

Discharge The volume of water flow through a given cross-section

Displacement A change in position of material point

Doline A funnel-shaped hollow in carbonate rocks. *Synonym: Sinkhole*

Elastic limit The point on a stress/strain curve at which the elastic behavior of the material transforms to inelastic behavior

Exposed karst Karstic rocks outcropping at the ground surface

Fault A fracture separating two bodies of rock, which have moved relatively to one another, or continue to move

Fault zone A zone with number of parallel or sub-parallel faults

Fissure Commonly used term in karst studies, is an open joint or crack, capable to provide a route for water movement and consequent rock dissolution

Fracture A more general term for all kinds of fracturing caused by mechanical stress in the rocks

Fracture pattern A special arrangement of a group of fracture surfaces

Fracturing A formation of breaks in a rock

Gallery A large, nearly horizontal passage in cave

Heterogeneity A variation of material properties from point to point

Homogeneity Material properties are identical everywhere

Isotropy Material properties are the same in all directions

Joint A fracture surface in the rock along which no displacement has occurred

Joint system See *Conjugated joints*

Karren The complex of superficial solution forms on compact carbonate rock. They vary in depth from a few millimeters to more than meter and are separated by ridges

Karst Internationally used term for specific morphological and hydrological features in soluble (mostly carbonate) rocks

Karst water A water dissolving the rocks during the passage through them and discharged from karst springs

Karstification A process (chemical and mechanical) of karst formation by water solution and infiltration

Laterite A tropical ferruginous clay soil

Lithology A physical characteristic and structure of rocks (composition, grain size, texture, degree of lithification)

Lithostratigraphy A system of formal naming and description on a local scale of defined recognizable units of rock successions

Normal fault A fault in which the lower block has moved downward relative to the upper block

Outcrop An open exposure of rocks, normally covered by younger sediments

Overthrust Thrust fault with a very low angle of dip and relatively large net of displacement

Permeability coefficient The rate of water flow through a unit cross-section and under unit hydraulic gradient

Phreatic cave A cave developed by dissolution bellow the water table

Plateau An elevated level and relatively flattened land surface

Polje A flat floored depression in karst areas whose long axis is parallel to major local structural trends

Ponor Hole in the bottom of a depression where a surface water stream flows underground into the karst system

Pothole Dominantly vertical cave or cave system

Precipice High vertical face of a rock

Reverse fault A fault where the hanging wall (lower block) has moved upward

Rock formation Lithostratigraphically distinct part of the lithosphere

Rupture The stage of the initial development of a fracture when instability occurs

Sedimentation The process of deposition of disintegrated rocks by water, wind, or gravitationally

Shear plane A plane along which of failure of materials occurs by shearing

Sinkhole The site where a surface water stream disappears underground

Slikenside Striated rock surface on a fault plane produced during the fault movement

Speleothem Term proposed by Moore (1952) for the unification of the secondary mineral deposits in caves

Spring, karst An emerging spring from karstified rocks

Spring, vauclusian The water rises under pressure from a vertical bedrock passage (Fontaine de Vaucluse, France)

Stress The average intensity of the strain applied on surface unit

Strike The azimuth of fault or joint plane

Stylolite An irregular suture-lake boundary along bedding planes in limestones. Their genesis is supposed to be provoked by dissolution under pressure

Terra-rossa Insoluble reddish-brown soil formed by rock weathering of limestones

Thrust A gently dipping fault where the lower block is moved upward

Underground river Water flowing trough caves, caverns and galleries in underground karst systems

MIX
Papier aus verantwortungsvollen Quellen
Paper from responsible sources
FSC® C105338

If you have any concerns about our products,
you can contact us on
ProductSafety@springernature.com

In case Publisher is established outside the EU,
the EU authorized representative is:
Springer Nature Customer Service Center GmbH
Europaplatz 3, 69115 Heidelberg, Germany

Printed by Libri Plureos GmbH
in Hamburg, Germany